SpringerBriefs in Food, Health, and Nutrition

Springer Briefs in Food, Health, and Nutrition present concise summaries of cutting edge research and practical applications across a wide range of topics related to the field of food science.

Editor-in-Chief
Richard W. Hartel
University of Wisconsin—Madison, USA

Associate Editors
J. Peter Clark, *Consultant to the Process Industries, USA*
John W. Finley, *Louisiana State University, USA*
David Rodriguez-Lazaro, *ITACyL, Spain*
David Topping, *CSIRO, Australia*

T0213819

For further volumes:
http://www.springer.com/series/10203

SpringerBriefs in Food, Health, and Nutrition

William Aspray • George Royer
Melissa G. Ocepek

Food in the Internet Age

 Springer

William Aspray
School of Information
University of Texas
Austin, TX, USA

George Royer
School of Information
University of Texas
Austin, TX, USA

Melissa G. Ocepek
School of Information
University of Texas
Austin, TX, USA

ISBN 978-3-319-01597-2   ISBN 978-3-319-01598-9 (eBook)
DOI 10.1007/978-3-319-01598-9
Springer Cham Heidelberg New York Dordrecht London

Library of Congress Control Number: 2013945302

Printed on acid-free paper

Springer is part of Springer Science+Business Media (www.springer.com)

Introduction

This brief book examines food in the Internet age. More specifically, it explores several ways in which the Internet is used by both individuals and organizations to carry out food-related tasks such as sharing recipes, reviewing restaurants, delivering groceries through online ordering, selling discount coupons, and providing cultural and political commentary on various aspects of food in everyday life. There is only brief discussion of the use of the Internet for business functions other than e-commerce, such as proprietary networks for internal communications, communication and ordering along supply chains, and online auctions of raw food products. While the book is primarily about food activities that take place in the United States, the lessons drawn about the business history of food companies, methods for building trust online, and harms to consumers and Main Street businesses by Internet companies apply internationally.

There is no attempt to be comprehensive in this book – a task that would be impossible in such a small study. Instead, we have selected four topics central to the Internet and food. Each chapter can be read as a stand-alone article, even though several of the topics carry across two or more of the chapters. The methods applied here are those of information studies and business history. Thus, in addition to telling about food in the Internet age, the book suggests ways in which food studies could be expanded by taking into greater consideration themes and approaches central to information studies and business history. The four chapters are described below.

How does one characterize the many different aspects of food online? It is a complex topic involving the production and selling of many different kinds of products and services, the sharing of information such as recipes and restaurant reviews, the critiquing of nutritional and cultural topics, and the expression of personal interests, activities, and identities. The first chapter provides four brief analyses that, together, provide an introduction to the various ways in which food appears online. The first analysis describes four categories of food websites according to a conceptual framework created by the authors. The second analysis uses data from the web analytics company Alexa to identify the most trafficked food websites in the United States, places these websites in groupings based on intended use, and shows how these food websites rank among all websites. The third analysis uses three case studies of websites created by individuals to illustrate the categories described in the first analysis. The fourth analysis provides a similar categorization of food applications for mobile phones. These four analyses provide a useful entry point to understanding the many different ways in which food appears online.

In 2008, when CNET published its list of the top ten dot-com flops ever, online grocer Webvan topped the list. The second chapter describes the rise and fall of Webvan and analyzes the reasons for its failure. In particular, this study demonstrates that Internet companies – contrary to what many entrepreneurs believed during the dot-com boom – are not immune to the basic laws of economics or sound business practice. Issues that are discussed include lack of knowledge of the grocery business, high efficiencies and low profit margins in the traditional grocery business, lack of consumer testing, inability to access a strong supplier market, cost of and problems with innovative and highly automated warehouses, the economics and logistics of home delivery, the "get big fast" philosophy common among dot-com entrepreneurs, the failure of other early online grocers such as Kozmo and UrbanFetch, and the growth of successful online grocery businesses by such companies as Peapod and Tesco.

When faced with the extraordinary successes of Internet companies such as Amazon and Google and the advantages they provide to both their customers and affiliated businesses, it is perhaps difficult to see the harm that Internet-based companies can create. The third chapter is intended to temper the sentiment about the virtues of the Internet. It examines three popular Internet-based, food-related companies and discusses how each has caused harm to either consumers or Main Street businesses. More specifically, the chapter studies the restaurant reservations company OpenTable and how consumer expectations that restaurants will add online reservation systems burden restaurant owners with high costs and may harm the ambience of restaurants; the online coupon company Groupon and how businesses can be harmed by spikes in demand and changed expectations of consumers; and the restaurant reviewing company Yelp and how it can ruin the surprise factor for new patrons, lead to hostilities between owners and reviewers, and harm restaurants through unsavory sales practices.

The final chapter focuses on trust online. The first section briefly considers some of the earliest examples of unfair online community reviewing, especially reviews that appeared on the website of the online retailer Amazon. The bulk of the chapter is focused on the trustworthiness of recipes that are posted and shared online, presenting six models used to make readers trust recipes they find online: the community review model as represented by Allrecipes; the laboratory testing model as represented by *Cook's Illustrated*; the scientific model as represented by Nathan Myrhvold, the Modernist Cuisine movement, and Alton Brown; the expert model as represented by the websites and blogs of ten professional chefs with well-trafficked websites; the corporate publishing model as represented by Epicurious, which reprints recipes formerly published in *Gourmet* or *Bon Appetit* magazines; and the corporate food products model as represented by the Betty Crocker brand of General Mills. The final section of the chapter considers the information studies literature on trust, and how that literature addresses issues of trust in both restaurant reviewing and recipe sharing.

Acknowledgments

This book is a product of the Informed Stomach Research Group in the School of Information at the University of Texas at Austin. We thank the School of Information for its infrastructural support in carrying out this and other research on food and information in the United States. William Aspray was supported in this research in part by research funds associated with the Bill and Lewis Suit Professorship in Information Technologies and the STARS Plus Fund provided by the University of Texas at Austin. Over the past 5 years, the Informed Stomach Research Group has received intellectual guidance from Harrison Archer, Zak Archer, Lecia Barker, Andrew Dillon, Lorrie Dong, Philip Doty, Elizabeth Englehardt, Nathan Ensmenger, Kenneth Fleischmann, Patricia Galloway, Barbara Hayes, Steve Mannheimer, Susan Smulyan, and Yan Zhang. Research assistance in archives, literature review and retrieval, and copyediting have been provided during this same period by Lynn Eaton, Kip Keller, and Elliot Williams. Thanks to all.

A Note on Usage

"This is the type of arrant pedantry up with which I will not put." (Winston Churchill, as quoted in goodreads.com)

Participants in the world of the Internet have taken license to be creative in the spelling and punctuation they use, both in the name of companies and websites, and in the ways in which they communicate in their online messages. One trait is to show excitement by the use of exclamation points (e.g., *Yelp!* or *Groupon Now!*). In this book, the authors have followed an increasingly common practice of dropping these exclamation points when the meaning is clear. Thus, in this text we use *Yelp*, not *Yelp!* Many companies use the *.com* suffix in their names. Also following increasingly common practice, in cases where it is clear we eliminate the suffix. So we write *Amazon* rather than *Amazon.com*. There are a few cases, however, where we retain the suffix for purposes of clarity, e.g., in Fig. 1.2 we write *Cooking.com* instead of *Cooking* because the latter could be confused with a statement about content when we meant the company of that name. There is no attempt to alter creative spelling, e.g., *Apartment Therapy's The Kitchn*. We also use neologisms where they are common, e.g., *advergames* and *pageviews*. Similarly, we don't mess with companies that run their names together, however they handle their capitalization, e.g., *OpenTable* or *Allrecipes*.

Contents

Chapter 1
Food Online: An Introduction to a Complex Environment

> *I want to be successful in everything I do, but the Internet and food just distract me. (Anonymous, Searchquotes 2013)*

1.1 Introduction

The World Wide Web is large. Nobody seems to know precisely how large, but estimates range from 15 billion to 50 billion websites worldwide (See www.worldwidewebsize.com). Beyond being large, the Internet is wonderfully diverse. Online content areas often reflect their offline popularity; as the culinary arts have become well represented on television networks, in magazines, and at the top of the *New York Times* bestsellers list, so food has taken over a large swath of the Internet. Cooking websites number at least 10,000 in the United States (based on Alexa's categorization of the 30 million websites it tracks – as discussed below), and this number is likely to be a gross underestimate. There are numerous other places where food is also found online – many of them associated with the businesses of producing, manufacturing, delivering, and selling food products in their many forms. To take just a few examples, food appears online in advertisements on many different kinds of websites, as an object for sale on e-commerce websites, and as a commodity to be exchanged or tracked on proprietary websites either internal to a company or used for interactions between a company and its suppliers. Food shows up online as a raw ingredient, a finished product, a resource, and a cultural artifact.

All of this is to say that the place of food online is a very complex topic, that it may be difficult to give a satisfactory accounting, and that any definitive accounting will necessarily be lengthy. We cannot hope to provide a definitive account of food online in the brief few pages allotted to this chapter. Instead, we provide four brief analyses, which together provide an introduction to the various ways in which food appears online. The first analysis describes four categories of food websites (referred to below as categories 1, 2, 3, and 4), according to a conceptual framework created by the authors. The second analysis uses data from the web analytics company

W. Aspray et al., *Food in the Internet Age*, SpringerBriefs in Food, Health, and Nutrition, DOI 10.1007/978-3-319-01598-9_1,
© William Aspray, George Royer, Melissa G. Ocepek 2013

Alexa to identify the most trafficked food websites in the United States. We add value to Alexa's rankings by grouping websites based on intended use, and show how these food websites rank among all websites. This analysis elaborates primarily on the category 3 and category 4 websites. The third analysis uses three case studies to elaborate on the category 1 and category 2 websites, and in particular to show how individuals use websites for personal and professional purposes. The fourth section complements the first three sections by providing a categorization of food applications for mobile phones. These four sections provide a useful entry point to understanding the many different ways in which food appears online.

1.2 A Conceptual Mapping of Food Online

Figure 1.1 maps the types of objects (blogs, podcasts, websites, etc.) related to food that one may encounter on the Internet and identifies both the actors responsible for their production and the intended audience. These actors are grouped into four major categories. This scheme employs concepts used commonly by business scholars. One could, of course, employ any number of alternative categorizations based, for example, on type of food or demographics of intended users.

The first category includes objects produced or encountered by individuals in the course of their daily lives that are shared through digital media with a limited audience, which typically includes only friends and family. Types of objects in this category include photos of food, videos of food preparation, recipes, personal websites about or including sections on food, blogs about or including entries on food, and posts on social media websites about food. These types of objects may be

	Category 1	Category 2	Category 3	Category 4
Actor	Individuals, sharing aspects of daily life	Individuals, motivated by production of "capital"	Product-oriented Business	Service-oriented Business
Audience	Friends and Family	Wide	Wide	Wide
Objects	Blogs, Websites, Social Media, Photos and Videos	Blogs, Websites, Social Media, Photos and Videos, Podcasts	Public Websites, Marketing Materials, E-Commerce Portals, Online Auctions, Company Proprietary Websites, Social Media Presence, Games	Public Websites, Mobile Applications, Marketing Materials, User Generated Reviews, Company Proprietary Websites

Fig. 1.1 Conceptual categorization of food objects on the Internet

understood as an individual's attempt to share aspects of her life with persons in her social or family circles. A distinguishing characteristic common to all materials falling in the first category is a lack of intention on the part of the object's creator to build any kind of capital (e.g., financial or social) through the production or sharing of food objects. This distinction is important because objects contained in the second and fourth categories may superficially resemble objects found in the first category. For example, while blogs may be found in multiple categories, blogs belonging to the first category are created for pleasure while those in the other categories are created in an attempt to gain revenue or establish oneself as a culinary expert.

The second category includes materials created by individuals with the intended purpose of gaining social, financial, political, or informational capital. Social capital may be accrued by networking with persons having similar food-related interests. For example, an individual could build her social capital by sharing recipes, critiquing restaurants, or maintaining a blog that demonstrates her mastery of a certain style of cooking. By accruing enough social capital, an individual may establish herself as an expert within the communities she wishes to be associated with. Occasionally, individuals can utilize their social capital to realize financial capital, e.g. revenue earned from advertisements on blogs or gifts of free meals provided to influential members of the food community. Where individuals demonstrate the power to shape discussions and influence the behaviors and opinions of players in the food industry, political capital may also reside. Lastly, individuals may accrue informational capital by virtue of their position in a community, gaining access to knowledge or skills possessed by experts or companies. For example, an individual who has demonstrated her knowledge of a particular food domain might be invited to participate in a proprietary disclosure of special techniques or secret recipes with expert chefs or food manufacturers.

Objects found within the second category include many of the same types of objects found in the first category – including social media postings, blogs, websites, photos, podcasts, and videos. The difference is that the materials are created with the intention that they will be consumed by a wider, possibly global audience and that the author intends for some form of demonstrable benefit to arise from the act of creating and sharing content. Objects found in the second category are much more likely to be food-specific than those in category one. For example, while a personal blog intended to share aspects of daily life might include some entries on food, it might also contain posts about pets, family, and hobbies. In contrast, a blog including food that is constructed with the intention to gain capital is likely to have a narrow focus on particular types of content. One would not expect to find information about remodeling a home, pictures of children or pets, or photos from a vacation on a category 2 website devoted to a specific style of cooking. Actors creating materials within this category include both professional and amateur bloggers, chefs and food critics, and professional food photographers.

Objects designed to create, promote, or distribute physical products fall within the third category. One type of object found in this category involves the production of food products by the food industries. They might take the form of a

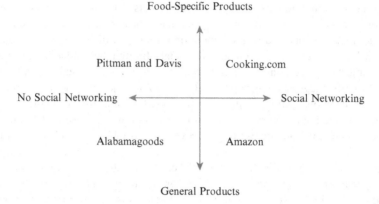

Fig. 1.2 Types of online product-oriented businesses

company-proprietary website for carrying out internal business within the company, a proprietary website that enables the company to work with its suppliers, or an online auction for the buying and selling of raw materials and finished products between businesses that may be geographically dispersed. We will not say much about these kinds of companies, but see Cortada (2008) for a general discussion of this topic.

Another class of online objects includes nutritional information, business locations, coupons, advertising material, photos or videos of products, web-based advergames, and materials promulgated through social media services such as Facebook and Twitter. All of these objects are directed at a general public and have a greater goal of selling more products, whether it is fast-food meals or artisan fruitcakes. As shown in Fig. 1.2, actors in this category are businesses that fall into one of four types depending on whether they are food-specific product vendors or general product vendors, and whether or not they employ social networking on their websites to carry out their sales. An example of a business that sells food-specific products but does not employ social networking is Pittman and Davis, a company that has sold fruits and gifts by mail since 1926. While customers may place orders online, Pittman and Davis does not enable customers to post comments, rate products, create lists, or engage in any type of social networking activity. The second type includes businesses that sell food-specific products and employ social networking. Cooking.com is an example. Its website, through which it sells a wide range of cooking-related products, invites users to rate products, search by product ratings, and post their own written reviews of products. The third type includes general retailers that sell products related to food (as well as many non-food products) and do not use social networking. Alabamagoods is an example. It is a specialty retailer that sells only goods manufactured in Alabama, including gourmet foods, barbeque sauce, jewelry, clothing, stationary, books, and pottery. Like Pittman and Davis, Alabamagoods's website does not allow product ratings or written reviews. The final type includes general retailers that sell products including food items and employ social networking. Amazon is an example. On Amazon's website, visitors

Fig. 1.3 Types of online service-oriented businesses

may purchase many types of product, rate and write reviews of products, search by ratings, and curate and share lists of products.

The fourth category includes objects created by business entities that function primarily to provide services. One type of website is devoted to internal company operations or interactions between the company and its suppliers, similar to what occurs in category 3. Other websites that fall in this category are concerned with the general public. As shown in Fig. 1.3, these websites fall into one of four types, depending on whether they provide food-specific services or general services that include food, and whether or not they employ social networking. Many, but not all of the services provided by this type of company are information services. An example of a business that provides food-specific services but does not utilize social networking is Zagat (as it operated prior to its acquisition by Google in 2011). Zagat provides an information service by rating and reviewing restaurants. In its pre-Google days, Zagat only offered its services online for a fee. After the acquisition, users who had a free Google + account could add content and access for free Zagat's full range of services, such as personalized restaurant recommendations. As such, Zagat as it is configured today is an interesting example of a service provider's adoption and integration of social networking into a traditional business. The second type includes businesses that provide food-specific services and employ social networking. Allrecipes, a website that provides information services in the form of recipes and cooking tips, is an example. Allrecipes enables users to rate both recipes and reviews, highlights blogs created by users, and provides a space for users to ask questions of their peers in Allrecipes's online community. The third type includes businesses that provide general services that have some relation to food, but are not only about food, and do not employ social networking. Groupon is an example. It provides its users with a primarily financial service by offering daily deals on a variety of services accessed through coupons purchased online. Discounts on meals at restaurants, delivered produce, and baked goods are among the types of services offered by Groupon, although Groupon also offers discounts on non-food items such as skydiving excursions and spa treatments. Groupon does not enable users to

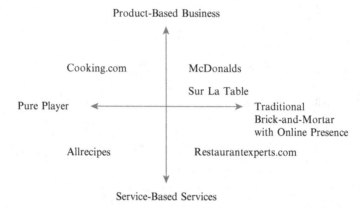

Fig. 1.4 Differences between pure players and traditional businesses with an online presence

write reviews, interact with one another, or create lists on its website. It does, however, pull ratings and reviews created by users on other websites such as Google, CitySearch, and Urbanspoon, to aid users in determining how desirable a given deal is. The fourth type includes businesses that provide general services and have some relation to food (but are not solely about food), but do employ social networking. Yelp is an example. It provides users with an information service in the form of a user-rated directory of local businesses. In addition to providing information about food-related businesses such as restaurants and grocery stores, Yelp also provides information regarding automobile services, home repair, health care, real estate, and financial services. Yelp relies on the content created by its active user community to populate its website – it is a heavily mediated environment in which users rate, recommend, and review primarily local businesses.

As detailed in Fig. 1.4, food-specific businesses may be further differentiated by considering the type of business model they employ. Both product and service businesses utilizing traditional brick-and-mortar business models with a web presence can be contrasted to the "pure players" who rely entirely on e-commerce for revenue. These distinctions break these companies into four categories. The first category comprises product-oriented businesses that are pure players. An example is Cooking.com. It sells thousands of food-related products, but has no physical stores. The second category consists of product-oriented businesses that maintain physical stores as well as have an online component. Many restaurant chains fit into this category, and McDonald's is a good example. On its website, McDonald's offers information services including menus, nutritional information, and restaurant locations. These services are intended to encourage consumers to visit McDonald's restaurants and purchase the Happy Meals from which it derives its profit. However, not all businesses belonging to this category rely solely on brick-and-mortar sales. Another example of a business in this category is Sur La Table. It sells kitchen gadgets in both physical and online locations. What characterizes businesses in the second category is that they maintain both physical and online presences. However, the nature of the online presence could be either informational (McDonald's) or about e-commerce (Sur La Table). Where the online presence is informational, the goal of the material is to convince the user to physically visit a bricks-and-mortar store to purchase products as opposed to purchasing them online. The third category includes service-oriented businesses that are "pure players." Allrecipes fits into this category, as it is an information service provider that has no physical presence. The fourth category includes traditional brick-and-mortar businesses that provide a service and have a presence online. Restaurantexperts is an example of such a business. It is the online home of Profit Line Consulting and offers business services to restaurants in the form of business consultation for a variety of types of restaurants. Through its website, it offers information resources and sells booklets on various aspects of the restaurant business. It also offers in-person and on-site consultation with its experts.

1.3 Popular Food Websites

In this section we provide a snapshot of the most visited food-related websites on the Internet by using data from a well-known web analytics company, Alexa: The Web Information Company (data taken from 17 May 2013). Alexa generates data for millions of websites on how often people visit each website (Global Web Traffic Estimates) and then ranks more than 30 million websites based on these estimates. Alexa displays its Top Sites ranking in two different ways: by category and by country. Alexa's cooking category (its only clearly defined food-related category) contains information on over 10,000 webpages. The cooking category is subdivided into 59 subcategories. Figure 1.5 represents our reorganization of Alexa's 59 subcategories into 10 conceptual groupings. The conceptual groupings are based on both the content of the websites that fall under each grouping and the likely ways in which visitors will use these websites. Whereas the previous section of this chapter studied the goals of the creators of the websites, this analysis instead concerns content and meaningfulness to users.

Alexa does not count every instance when someone clicks on a particular website. Instead it provides browser extensions that are installed on the computers that appear on the desks of individual Internet users. When installed, these browser extensions automatically collect and record the webpages where these computers browse. The number of computers having the Alexa browser extension installed numbers in the millions worldwide, so Alexa has some confidence that its data reflects actual Internet traffic patterns reasonably well. Alexa's global traffic rank is a measure of a website's traffic relative to all other websites tracked by Alexa over the past three months, based on its own data. The calculations involve a combination of the estimated averaged daily unique visitors to the website and the estimated number of "pageviews" on the website over the past 3 months. Low-traffic websites (i.e. websites not among the 100,000 most heavily visited websites) receive only rough

Conceptual Groupings	Alexa Subcategories	Total Number of Webpages
Dish types	Soups and Stews (2,561)	8267
	World Cuisines (1,425)	
	Fruits and Vegetables (1,083)	
	Sandwiches (489)	
	Fish and Seafood (427)	
	Meat (380)	
	Beverages (378)	
	Holidays (262)	
	Condiments (106)	
	Nuts and Seeds (103)	
	Salads (101)	
	Sauces, Dips, Gravies, and Toppings (93)	
	Grains (92)	
	Desserts (84)	
	Cheese (73)	
	Pasta (70)	
	Eggs (63)	
	Herbs and Spices (59)	
	Appetizers (55)	
	Jams, Jellies, and Preserves (50)	
	Pizza (45)	
	Chili (42)	
	Dairy (39)	
	Breakfast (37)	
	Casseroles (29)	
	Gourmet (21)	
	Stuffings and Dressings (20)	
	Wild Foods (18)	
	Chocolate (16)	
	Meat Substitutes (13)	
	Snacks (13)	
	Spicy (13)	
	Fusion (7)	
Techniques	Baking and Confections (635)	878
	Crock-Pot (44)	
	Canning and Freezing (36)	
	Microwave (31)	
	Techniques (28)	
	Gifts in a Jar (25)	
	Fondue (18)	
	Quick and Easy (16)	
	Drying and Dehydrating (15)	
	Pressure Cooker (15)	
	Quantity Cooking (15)	
Dietary Needs	Vegetarian (163)	398
	Special Diets (143)	
	For Children (92)	
Recipe Collections	Recipe Collections (386)	386
Food Information Venues	Weblogs (314)	366
	Mailing Lists (31)	
	Chats and Forums (21)	
Brand Name Company/Celebrity Chef Created Recipes	Brand Name Recipes (102)	204
	Chefs (102)	
Periodicals	Magazines and E-zines (40)	92
Food Activities	Outdoors (59)	71
	Tailgating and Picnics (12)	
Food Safety	Safety (20)	20
Personal/Organizational Information Management	Recipe Management (15)	15

Fig. 1.5 Conceptual grouping of cooking websites (Source: Based on the cooking top sites world-wide provided by Alexa as of 17 May 2013. The *numbers in parentheses* are the number of websites identified in that particular Alexa subcategory)

estimates. We do not have sufficient information to determine how closely Alexa's numbers match actual Internet traffic, but for the basic conceptual grouping we do here, Alexa's large sample size makes it clearly good enough for our purposes.

Some final methodological points before turning to the analysis of Fig. 1.5: When deciding in which conceptual grouping to place each Alexa cooking subcategory, we analyzed several of the websites listed within the subcategory. We placed each subcategory into exactly one conceptual grouping. The fit was not always perfect. For example, Baking and Confections, Quick and Easy, and Fondue were placed in the Technique conceptual grouping, but they also contain webpages that could have been placed in Dish Type. Figure 1.5 refers to webpages because Alexa organizes its categories by webpage, whereas the Global Web Traffic Estimates metric ranks websites and does not provide analytics for specific webpages within a website.

Any Alexa subcategory that references a specific food, meal, or cuisine we place in the Dish Type conceptual grouping. Most of the cooking websites fall under Dish Type. These webpages primarily include recipes and ingredient information. Of all the Alexa subcategories we list under Dish Type, Soups and Stews represents the largest number of websites. The top Soups and Stews webpages include, for example, both the Allrecipes Soup page as well as several Allrecipes pages dedicated to specific types of soups, such as baked potato soup and salmon chowder.

Several other conceptual groupings also refer to webpages primarily focused on recipes. The second most heavily populated conceptual grouping is Technique. Alexa subcategories referencing a specific cooking method that does not refer to a finished food product are included in the Technique conceptual grouping. The webpages here primarily consist of recipes focused on a specific technique applied to a range of ingredients. For example, Microwave Cooking for One is a webpage focused on cooking solely with a microwave. The third most heavily populated conceptual grouping is Dietary Needs. It comprises webpages specializing in recipes tailored to dietary restrictions or specific audiences such as children. For example, the Alexa subcategory Special Diets is where one finds webpages for low-fat, gluten-free, low-carb, diabetic-friendly, and several other types of diet-specific recipes. The conceptual grouping for Brand Name Company/Celebrity Chef Created Recipes and Food Activities also mostly contains recipe webpages.

The other conceptual groupings concern either food safety or information sources/resources for organizing cooking and recipe information. The conceptual grouping for Recipe Collections contains webpages that offer a wide range of recipes for users to search or browse. The main Allrecipes page is the most visited webpage in this category. The conceptual grouping for Food Information Venues contains a variety of webpages for users to access cooking information beyond recipe collections or specific recipe websites. Blogs, mailing lists, and message boards populate this category and, although much of the information contained within these venues involves recipes, the user accesses them in a different way or on pages that often offer additional content. The Periodicals conceptual grouping includes both online-only e-zines and magazines that have both a physical, paper embodiment and a presence online. For example, *Epicurus* is an e-zine that presents recipes

Fig. 1.6 Top 10 US Cooking websites (Source: Based on the Cooking top sites worldwide and then resorted by US rankings provided by Alexa as of 17 May 2013. The rankings were created by reviewing the US rankings of the top 20 cooking websites worldwide and the top 500 websites (of every kind) in the United States)

Rank	Website	Overall US Ranking
1	Allrecipes	184
2	Food.com	410
3	Recipesfinder.com	494
4	Cooks.com	1053
5	Epicurious.com	1106
6	Betty Crocker	1347
7	Apartment Therapy's The Kitchn	1581
8	Kraft Food Recipes	1766
9	EatingWell	1791
10	The Pillsbury Company	1844

Fig. 1.7 Top food-related websites in the US (Source: Based on the 500 top sites in the US provided by Alexa as of 17 May 2013. An *asterisk* (*) denotes websites that have a high percentage of food content but that also have non-food content)

Food Related Website	Overall US Ranking
Yelp*	48
Groupon*	86
Allrecipes	184
Coupons.com*	190
Livingsocial*	202
Food Network	232
Costco*	238
Sam's Club*	372
Food.com	410
WeightWatchers.com	491
Recipesfinder.com	494

and articles free online, while *Food and Wine* is a traditional print magazine that in its online version provides additional content beyond the print publication. The Personal/Organizational Information Management conceptual grouping includes pages involving software for organizing recipe collections, creating shopping lists, and hosting websites.

Figures 1.6 and 1.7 are also based on an analysis of the Alexa data. The focus here is not on the largest categories of food-related websites, as before, but instead on which particular food-related websites are most heavily trafficked. Figure 1.6 presents the top 10 cooking websites in the United States as recorded by Alexa. All of these websites include information about recipes, food products, or tools used in cooking. All of them were classified by Alexa as cooking websites. The majority of these top 10 websites each contains recipe collections or provide recipes from brand name food companies. Apartment Therapy's The Kitchn is unique to this list since it is primarily a daily blog with articles, product reviews, kitchen design, and recipes.

Figure 1.7 includes a wider class of food-related websites than is listed in Fig. 1.6, including ones that provide restaurant reviews, coupons, and weight-watching advice and support. The range of the websites listed here suggests the great variety of food-related websites that appear online. For example, there are websites in this list that offer restaurant deals, provide grocery coupons, or represent retail stores that sell a large volume of food products. The websites marked with an asterisk are primarily but not entirely concerned with food. For example, Yelp reviews bookstores, but it reviews many more restaurants; Costco sells clothing, but its highest volume come from selling food products. Entries not marked with an asterisk are totally devoted to food issues.

1.4 Understanding Personal and Professional Uses of Food Websites by Individuals

This section presents three case studies to elaborate on the categories described in the first section, and in particular to illustrate how individuals use websites for personal and professional purposes. The first case study is of Mrs. Q (Sarah Wu), who begins to blog for her own personal reasons (category 1), but then her work evolves into a category 2 website intended to build political capital and written for a general public concerned about nutritional issues associated with public school lunches. The second case study is of the Yelp Elite Squad and, in particular one Yelp Elite member, Kelly "Reverend" S. (real name Kelly Stocker), who is the Senior Community Manager for Yelp in Austin, Texas. This case illustrates how an individual can use a category 4 website (that of the online reviewing company Yelp) as an infrastructure to carry out the category 2 activity of gaining social capital. The third case study is of Joy the Baker (Joy Wilson), a California-based blogger who moves from category 2 to category 3 by building a business out of a cooking hobby.

1.4.1 Case 1: Mrs. Q (Sarah Wu)

Let us first consider the case of Sarah Wu, a speech pathologist who taught in a public elementary school in Chicago, who has become an advocate for better school lunches. We will tell the story from well before she began her blog, *Fed Up With Lunch*, because it shows the way in which a person evolves from category 1 (a web intended for friends and family) to category 2 (one for a public with a political purpose). We will tell the story as much as possible in Wu's own words.

As with many people, Wu began her blogging life (in 2004) as a category 1 blogger, writing about personal matters to families and friends. She continued to update this blog into 2009.

> I should tell you that I have a personal blog that I started about six years ago. My most consistent readers were my mother, my sister, and the friend who introduced me to the concept of a blog. I wrote about lots of personal stuff under an alias (some things never change, eh?). Oddly never about food, but about my family, my work, my life. I was lucky if I got 10 pageviews a week. But that was fine with me: I was not writing for anyone but myself and my family & friends.

> ... Going back to 2004 when a friend told me about her blog, I thought it was great. Every day I checked her blog to see what she was thinking about and doing. She suggested I start my own personal blog so I did. Weirdly, she ended up stopping blogging two years later, but I found it to be a great outlet for me and kept it up until I had a baby and then I only blogged about once a month.
>
> [After four paragraphs talking about how she kept a journal as a child, she continues about blogging.] Thankfully that friend introduced me to blogging. I found the experience to be very different than writing in a journal. The pros to blogging are that it's much easier to see how you change over time, it's easy to share, it's searchable, and it's fun to do something online. (Mrs. Q 2010c)

One day in 2009, Wu had forgotten her lunch and ate in the school cafeteria. She was appalled by the food: a plastic-wrapped Barkin' Bagel (a tasteless hot dog in a soggy bun, cooked and served in a shrink-wrapped package), six tater tots (the mandated vegetable), a few cubes of pear suspended in red gelatin, and chocolate milk. This experience created in Wu a desire to understand and share what public school lunches are really like. She decided to eat in the school cafeteria every day for a year, photograph the meals with her cell phone, and blog about her experience. She was particularly concerned because over 90 % of the kids in her school were able to eat lunch only because they received free or subsidized meals through the federal lunch program. As she noted at the beginning of her project:

> It's very challenging to teach students when they are eating school lunches that don't give them the nutrition they need and deserve. Oftentimes what is served barely passes muster as something edible. And after a meal high in sugar and fat and low in fiber, they then must pay attention in a classroom.
>
> I'm going to attempt to eat school lunch everyday in 2010. As a teacher it's available to me as well for a few dollars. Most of the students at my school get free lunch or reduced. I'm going to take pictures of the school lunch and post them.
>
> Normally I shop for organic fruits and veggies. I avoid processed foods and food high in sugar and high fructose corn syrup, but I normally eat food brought from hom[e] including leftovers, sandwich, or a "healthy" microwave meal for lunch with yogurt and a piece of fruit.
>
> The reason I'm eating school lunch every day is to raise awareness about school lunch food in America. (Mrs. Q 2010a)

Note in the quotation below that when she began writing her school food blog, Wu was doing so on a "whim". She was not sure that it would have traction with a larger audience, but she wanted to write something "meaningful". This might still be considered a category 1 web activity because she did not have an expectation of a public audience for what she was writing, but when she found that there was a public interest her blog clearly transformed into a category 2 activity, intended for political action and directed at a public audience. In an interview on television's Good Morning America (October 5, 2011) she said, "I never intended for it [the blog] to get the attention it did" (Zhao 2011). However, when the blog was only 2 weeks old it started to attract attention, and over the entire year of 2010 the blog received more than a million hits.

> So just so it's clear, Fed Up With Lunch didn't come out of a vacuum. But the decision to blog my journey eating school lunch every day was basically made on a whim. Late in 2009 I thought about other people doing yearlong projects (mostly the 365 photo blogs) and I liked that idea, but wanted something a little more meaningful than just random photos of

my life. I thought about my work and my students and this idea literally just came to me. POP! Fed Up With Lunch was born. (I'm actually not religious, but I gotta wonder about a higher power…). (Mrs. Q 2010c)

Wu's use of Twitter is another clear sign that she had transformed from a personal to a public goal. Her first exposure to Twitter was personal ("it did not seem my thing") but then she changed her mind, again for personal reasons (to follow some famous people) and opened a protected, personal Twitter account. But when she saw how Twitter could work in tandem with her school food blog, she had clearly transformed from Category 1 to Category 2.

Well, that friend [who introduced me to blogging in 2004] is the exact same person who introduced me to Twitter. She got an account pretty early on. I looked around on Twitter after she told me about it. It didn't seem like it was my thing so I didn't get an account at first. I didn't want to give status updates to people I didn't know. But after a couple months, I figured I would get an account too, mostly because I wanted to follow people who were famous (Mark Bittman, Rick Bayless, Yoko Ono, Venus Williams, Andy Roddick, Al Gore, Dooce, Anderson Cooper (side note: yum), Augusten Burroughs, Oprah, and on…) Yes, I have a personal Twitter account that I used for a short while with about 150 followers. It's a protected acct, but don't worry: I no longer tweet there (too busy for two accounts).

I started a Twitter account for Fed Up With Lunch because I liked that Twitter had a broader scope than Facebook, but it was more "anonymous." Since I am anonymous, that feature lined right up with my initial goals. I thought about starting a Facebook account for this project, but I didn't want it to be found by my coworkers who are on Facebook. Although there are tons of people on Twitter, only 2 % of my real life friends are on Twitter compared to 98 % of them are on Facebook.

Now I use Twitter more than I use Facebook. I still don't have an official Facebook account for the blog project (but there are two imposter ones—shaking my fist at weird strangers!). I'll probably start up a Facebook account for the blog, but I'm still in no hurry.

I enjoy tweeting because it's fun to connect with people who are interested in similar topics (food, education, school lunch, parenting, etc). They are people I would never have met without having a Twitter account. Also I credit my Twitter account for making Fed Up With Lunch a successful blog. It's the PR department! (Mrs. Q 2010d)

The entries in the Fed Up With Lunch blog were posted almost every day. Wu discussed what she had eaten that day and provided color photos of the meals. She often remarked on how bland and tasteless the food was, how much processed food was served, and how high the salt content was. She would also often discuss her physical or mental reaction to the food. Here is a typical day's entry:

Day 80: Meatloaf
Today's menu: meatloaf, mystery greens, cornbread muffin, pineapple fruit cup, milk
Mystery greens are back! They weren't as bitter this time: I didn't have to spit them out. But I still couldn't finish more than a couple bites. Yuck.
The meatloaf looks weird, no? It's yellowish in color. It tasted like a salty commercial hamburger patty. It sounds disgusting, but I sopped up the watery gravy with the cornbread muffin. I felt sick about an hour and a half after eating lunch. I thought I was going to throw up. But thankfully it passed and I went on with the day. No students complained so it must have been just me.
I asked a student, "What did we eat for lunch?" He replied, "Chicken". (Mrs. Q 2010b)

Wu wrote the blog anonymously, under the pseudonym Mrs. Q, because she was afraid that she would be fired by the school district if her identity were revealed.

When a reporter from the *Chicago Tribune* wanted to interview her in 2010, she got cold feet. She was worried that "a *Chicago Tribune* article could have resulted in a witch hunt instructing all lunch room managers to report back on teachers who were regularly buying lunch." The reporter inquired "with the Chicago Public Schools about the district's potential response should, say, any of the systems decide to take a picture of his/her own lunch and post it online. Representatives would only say that lawyers for the Chicago Teachers Union would have to counsel that teacher. That seemed frightening enough" (Eng 2011). When asked why she decided to give up her anonymity, Wu replied:

> Around halfway through the project, I was approached by a literary agent about turning the blog and the experience of eating school lunch into a book. I talked with my agent about it and she believed that an anonymous book would not sell. I had to make a choice. Did I want to write a book and reveal myself or pass and stay anonymous? Ultimately, I decided that I wanted information about school lunch reform to reach a wider audience and so I decided to write a book. (D'Arcy 2011)

In 2011 Wu published her book, entitled *Fed Up with Lunch: The School Lunch Project: How One Anonymous Teacher Revealed the Truth About School Lunches – And How We Can Change Them!*, which pulled together material from her blog. At that time, she revealed her identity (D'Arcy 2011; Dell'Antonia 2011). She did not lose her job, but in 2012 she resigned from the Chicago Public Schools to spend more time with her family, given the long commute to work and a new baby in the family. While the blog still exists, it has not been actively updated with food-related material since October 2012, just before the new baby arrived. Many of the entries of 2013 are baby photos, indicating the use of the blog more as a personal than a professional website in recent times.

However, there does continue to be an active professional *Fed Up With Lunch* website on Facebook. The Facebook website has 4,710 people who "like" it (as of May 31, 2013). It includes posts by Wu and others not only about school lunches, but also about other food topics such as eating insects to fight obesity and links to a YouTube video of Michael Pollan going grocery shopping. There is still a political edge to the Facebook website, e.g. an entry supporting an Indiana farmer who challenged and lost in court in his bid to use seeds from soybean plants that had been grown from Monsanto seeds, and encouraging readers to write to the USDA about their concerns.

1.4.2   Case 2: Yelp Elite Squad and Kelly "Reverend" S. (Kelly Stocker)

As will be discussed in Chap. 3, the social networking-centered reviewing company Yelp has a business model that involves building in each of its principal places of operation a cadre of reviewers, known as the Yelp Elite Squad. These are people who the company relies upon to write reviews in large numbers and of wide interest. The company pampers its Elite members by providing them with various types of acknowledgment online and by inviting them to exclusive events such as free dinners and cocktail parties, which are generally underwritten by local businesses interested in attracting Yelp members to be their customers. From this practice the company gains an active and enthusiastic reviewer base used to ensure both thorough coverage of local businesses and a buzz around Yelp. In establishing a profile strong enough to be invited to be an Elite member, it is not uncommon for a member to spend 20 h per week of her own time reviewing for Yelp (Owyang 2008).

Attaining membership in the Yelp Elite Squad is a way for individuals to use a Category 4 website to build their own social capital. Viewed from the perspective of the individual rather than the perspective of the company, this is a case of category 2, an individual using an online presence to create social capital for herself.

Yelp Elite membership is considered by many food enthusiasts to be valuable social capital. Consider the case of Yelp member Vyvy "Batman" D. In November 2012, Ryan C., the Yelp Community Manager of Orange County, California, wrote to invite Vyvy D. to be a Yelp Elite member:

> Not to come off as a creep, but I've had my eye on you lately. In fact, the entire Yelp Elite Council has. You clearly embody the spirit of Yelp with your enthusiasm, positivity, constructive honesty and useful funny cool-ness, and we'd like to formally invite you to join the Yelp Elite Squad! As a member of the best and brightest on Yelp, you'll be invited to exclusive parties, where you'll rub elbows with some of our city's most influential movers and shakers. Plus, you'll get a snazzy badge on your profile so everyone can see you know what's up! (Vyvacious, November 20, 2012)

Vyvy D. expressed her elation on her personal blog: "I'm Yelp Elite, BITCHESSS!!!" and she goes on to say, "I don't think you understand how *ecstatic* I am…well maybe you have some idea but really…I'm so freaking excited." This is a woman who sizzles with emotion from having gained social capital through Yelp. Or consider the more muted reflections of blogger Weiward Girl:

> I Yelp. I've been doing it since October 2007. At first, I *used* Yelp. Our relationship was casual. There are so many great reviews and pictures from actual diners, people who share my love of good food and drinks. I searched and read and discovered great new eateries. As I dined around town, I was eager to write more reviews of my own, both good and bad. I wanted other diners to hear about my experiences. It became increasingly fun to get and give compliments on reviews, and I kept clicking the radio buttons to rate them: "useful, funny, cool." Soon, friend requests starting showing up, and that's when I realized: I've become a real Yelper! There was nothing stopping me from going all out now. My new Yelp friends were hanging out in the talk threads and I joined them. The camaraderie felt real. All the regular Yelpers are really in the know about all things fun and tasty around the city. Before long, a Yelp ambassador sent me a message: "Would you like to become an elite Yelper?" Boy, would I ever! I must have been doing something right to get such recognition so soon! Being an elite Yelper means a shiny new badge on my profile, and invitations to exclusive parties at the hottest spots in town! Just days after my promotion to the Yelp Elite status, I went to a party at a hip lounge. There were free-flowing vodka martinis. I was impressed. Yelp must be quite influential and have a generous budget to pull this off! So many people welcomed me. "They couldn't wait to meet you!" the Yelp ambassador told me as soon as I entered. The music was loud. I was shoulder-to-shoulder with elite Yelpers. Cameras were flashing. Cocktail glasses were clinking. Everyone was laughing, hugging and having a good time. Nobody remembered it was a school night. I bet the lounge that hosted the party will be getting a lot more new business now that all the Yelp elites have been there. Good for them! (Weiward Girl 2009)

Occasionally, but not often, the social capital acquired through Yelp translates into some kind of employment. For example, Ed Uyeshima obtained a job as a travel writer for the San Francisco Examiner, Chris Hansen found a job working for the influential website midtown.lunch.com, and a few people (such as Kelly S., who is profiled below) were hired by Yelp, typically as community managers to organize the Elite events and bring cohesion to the Yelp Elite membership in a particular metropolitan area. For the most part, bootstrapping a career through active Yelp participation is as elusive as the goal of schoolyard basketball players to earn a spot on an NBA team. More commonly, the payoff of active participation in Yelp comes in the form of social contacts and heightened respect:

> For others, the rewards are personal. San Francisco architect and amateur pickled meats expert Theodore Ordon-Yausi, 26, was tapped in 2010 after attending the "Elite Prom" as a plus-one. "My reviews are definitely read more now," says Ordon-Yaussi, who writes under the nom de plume The King of Pastrami. "I get more random messages from people I don't know. I have 20 new fans who follow my reviews." That can be invigorating. "I like to tell my friends that my opinion is important," says 29-year-old research scientist Kristin Patrick. "When I go into a restaurant, the owner says, 'You're here to write a review, aren't you?' I'll say, 'Don't worry, I'm Elite.'" (Sax 2011)

What happens at one of these Yelp Elite events? Journalist Molly McHugh of *Digital Trends* was taken along to one of these exclusive events by some of her friends who were Yelp Elite members. Here is an excerpt from her report on the event:

> Let me tell you, it's all a little surreal. There's a line where you check in, ID in hand, and are given a name tag to wear. After making it past this very relaxed gate-keeping, there's nothing left to do but party. Allow me to set the scene:
>
> The event is in a hip ballroom in the cool kids' part of town. In the middle of the room, there are about five tables featuring local restaurants who are handing out free plates of their food. There are tamales, gourmet ice cream, deli sandwiches, flan, bread pudding, a sausage-deviled egg item, and a few other appetizer-type plates.
>
> And then – and then! – there's the bar. There are four or five specialty cocktails for the event, and since I live in Portland, Oregon, there are six or so beers on tap. A sparkling wine from a local winery is also featured. I try many of these; the lines never get too long, the bartenders always cheerful, the booze continuously flowing. Trust us – Yelp Elite know how to drink….
>
> Now, we journey upstairs. There's a row of tables doling out samples: Yet more wine, as well as vodka. Yes, "sampling" vodka is only mildly less college than doing a shot. Yes, they're basically the same thing. At the end of the row, a line is forming in another room for free 10 minute massages. Another station is a local barber shop offering straight-razor neck shaves and hand massages. There's a photo area where there's a backdrop, professional lighting, and props for official Yelp event photos. Tables everywhere are covered with free swag for you to take: Yelp bottle openers, Yelp fingerless gloves, Yelp fans. (McHugh 2013)

What is the work that a Yelp member does to achieve Elite status? One can understand this by profiling Kelly "Reverend" S. (real name Kelly Stocker), who is the Senior Community Manager for Yelp Austin. One might argue that Kelly S. is not a representative choice because she is not only a Yelp Elite member but also a Yelp employee. However, many community managers begin as Yelp volunteers (See, for example, the stories of two other community managers: Susan 2009;

Rachel 2011). And as Kelly S. indicates, becoming an employee was just a continuation of her earlier life as an Elite member:

> Still rockin' Austin's Yelp scene... just with a fancier title and some hot projects. Writing, planning parties, social media-izing, rubbing elbows, kicking ass and kissing babies but working more strategically on projects and processes that affect the Community team on a larger scale. (Stocker n.d.)

In her About Me profile, Kelly S. lists herself as a Yelp evangelist, party planner, business lover, and spider monkey. She has been an Elite member every year she has been active on Yelp. Using the standard Yelp identity characteristics, Stocker identifies herself (among other ways) as being from the hometown of McAllen, Texas, a fan of the music from Johnny D and the Rocket 88s, and partial to gifts of flowers from men. She graduated from Notre Dame with bachelor degrees in management information science and English. Her tagline reads: "From A+on my report cards to writing about bars. My parents are so proud." All of this gives one a sense of who Kelly S. is without having to actually get to know her. The inclusion of this type of information differentiates Kelly S. from many category 2 actors, as it includes details about aspects of life beyond food. It also differs in that Yelp intends to let the reader get to know the Elite member without identifying exactly who she is, but Stocker has chosen to reveal her identity by giving her last name.

Kelly S. began Yelping in March 2009, and since then she has been a busy woman. She has created 1,105 reviews and 148 review updates. She has published 92 firsts (that is, the first review on Yelp of a particular establishment), 652 tips (such as suggesting one experience the *verde* sauce at a particular Mexican restaurant), 973 events (such as the LoneStar Crawfish Festival being held on a particular date at a local barbecue restaurant), and posted 4,234 local photos (such as 13 recent ones of a particular local food truck and its products). Her 259 lists include, for example, lists of 46 dive bars, 21 places "worth the trip", 33 places to go for WiFi, and 13 Austin "institutions".

Her reviews have been well regarded by Yelp members. 8,501 Yelpers have found her reviews useful, 5,813 have found them funny, and 6,697 have found them cool. She has also received numerous compliments from other Yelpers, for example 2,366 people have thanked her, 1,014 people have said she is "cool", 555 have said she is "hot stuff", 539 have said she is funny, and 338 have said she is a good writer.

The Yelp Elite Squad has attracted some critics. For example, Amy Blair has ridiculed the inanity of the reviews written of a Yelp Elite event held at the recently opened Four Seasons Hotel in Silicon Valley (Balla 2008). Others have criticized the "cultish" nature of the Yelp Elite meetings and have complained about the manners of the Yelp Elite members at these events, likening the "rabid ferocity with which certain guests attacked...[the] hors d'ouerves [at a Yelp Elite meeting in New York City]" to "an Animal Planet feeding frenzy" (comments of Yelp's New York community manager as quoted in Sax 2011; for a similar criticism, see Chen 2011). Several people have complained about the audacity of Yelp Elite members in expecting businesses to cater to them. For example, even before their opening day in business, an Elite Yelper wrote to a Big Gay Ice Cream company's retail outlet, pressuring them to open early one particular day, prepare samples of every flavor for everyone in their entourage, and create a special flavor of the day so that a group of Elite Yelpers could go on an ice cream crawl (Nguyen 2012). The apotheosis of chutzpah is the ReviewerCard, created by 35-year old Brad Newman, which is to be shown to proprietors to pressure them to provide discounts, free products and services, or other perks to holders of the card in exchange for writing positive reviews or restraining from writing negative reviews (Beck 2013).

While we have focused on the Yelp Elite Squad here, it is not uncommon to find category 2 actors generating social capital by creating content on Websites operated by category 3 or category 4 businesses. It is likewise common to find category 3 or category 4 businesses cultivating community by offering rewards to top contributors. Amazon, for example, also recognizes its elite members. It offers "Hall of Fame" status for reviewers who rise to a high level in reviewer rankings. The Hall of Fame inductees receive a permanent Hall of Fame Reviewer Badge and are listed on a Hall of Fame Webpage.

1.4.3   Case 3: Joy the Baker (Joy Wilson)

Joy Wilson is a California-based blogger behind Joy the Baker. Her whimsical, down-to-earth, and humorous food blog quickly established her as one of the Web's most celebrated food bloggers. Examining the case of Joy Wilson, we observe that it details a particular kind of category 2 actor who becomes a category 3 actor by building a business out of a cooking hobby. Wilson began writing her blog in January of 2008 (Mullen 2012). Rather than merely provide recipes, Wilson's writings place her cooking within the context of her life. In a Los Angeles Times interview, Wilson (2011) explains that "There is a voice beyond the food. I express a lightness and vulnerability in my writing. My readers can relate to me, whether I break up with a boy, am going to a party, or a trip to the market." (10/7/2011). Joy the Baker has received numerous awards, including being named one of the eight best food bloggers by *Forbes* and one of the top 50 food blogs in the world by *The London Times* (Forbes 2010). The success of her blog enabled Wilson to secure a publishing deal for her own cookbook, and she embarked on a book tour in 2012. Although she took her blogging seriously, financial gain was not the reason for creating the blog. As Wilson explains:

> I feel like an accidental business lady. Joy the Baker was born out of this absolute love for food and baking, and I never anticipated that it would become a full-blown business. Because I don't always feel like a natural business lady, the business end feels overwhelming at times. (Mullen 2012)

Wilson began blogging as a way to participate and contribute to the online food community. Inspired by the materials posted by popular food blogs, Wilson decided that she would start a blog of her own.

I started Joy the Baker in January of 2008. I had just spent far too many years in college (I graduated when I was 25) and was working as a professional baker and personal assistant. I started Joy the Baker as a way to document all of the things that were coming out of my kitchen. My intentions in starting the blog were somewhat haphazard; I was driven by an unreasonable need to document my food and recipes—I just wanted to share it all. From Day One, I took my blog very seriously, but I took it seriously in my own head. I was always trying to learn more, take prettier pictures, make more delicious recipes, and generally outdo myself. For that reason, my blog was more than just a hobby for me, it was the space I had created for myself to push myself forward. (Mullen 2012)

Wilson's diligent blogging paid off. Financial opportunities appeared as her popularity within the food community increased. She credits her success to her ability to connect with others in the online community. Social networking technologies played an important role in Wilson's rise. In her own words:

When I started my blog I decided that I was going to completely immerse myself in the community. I joined a few baking groups online, and I would comment on fellow bloggers' work nonstop. Nothing I did was specifically designed to build my traffic. I was genuinely interested in the work other people were creating. I think that becoming an enthusiastic and vocal part of a growing community helped build my site over time. (Mullen 2012)

Wilson's case has some similarities with those of other category 3 and 4 websites. Allrecipes, for example, grew out of the founder's desire to collect and share cookie recipes with individuals seeking cookie recipes online. In Wilson's case, the desire to share information, not only about cooking but also about her life, eventually led to enough financial capital to allow her to make her hobby into her business.

1.5 Popular Food Apps

In this section we take a snapshot (as of 6 February 2012) of mobile phone applications ("apps") related to food. Rather than give a comprehensive portrait of food apps, we have analyzed here the top 100 mobile phone apps for food on the Android operating system (mobile phone platform), ranked by relevance. There is significant duplication of food applications on different mobile phone platforms, so there would be limited value in examining multiple platforms. Android is the platform with the largest number of apps, with close competition from Apple's iOS platform. The 100 most relevant applications are represented in Fig. 1.8. They are classified into types of apps, most of which are discussed below. Note that these apps are ones that are intended for use by individual customers. None of the mobile phone apps in the top 100 are intended for use in business-to-business applications.

The largest number of applications were designed for watching one's weight, mainly through counting calories and sometimes also through counting nutrients. Of these, many were designed for calculating Weight Watchers points, not only because Weight Watchers is the largest diet organization in the United States, but also because its system allows dieters to eat any food, so long as they note the points

Conceptual Grouping	#Apps in Grouping
Diet, calorie counter, nutrition counter	24
Restaurant finders	17
Recipes	15
Games	11
Coupons and discounts	7
Healthy eating, organics	7
Meal and food planners	4
Learn food terms in other languages	4
Meal delivery	3
Wine advice	2
Food education	2
Cooking school	1
Mobile access to food website	1
Mobile access to television (Food Network)	1

Fig. 1.8 Hundred top food applications in the Android marketplace (Source: Based on the top 100 applications (free or for cost), as of 6 February 2012, using "food" as the search term, searching market.android.com, ordering the results by relevance (not by popularity). We sampled 105 items since five results were spurious, e.g. Musical Soul Food is a gospel radio station, not a food-related application)

associated with it. Thus, dieting is an information-rich activity when one follows the Weight Watchers system, one that leads to numerous Android applications, such as the WW Points Plus Diary & Scanner.

Perhaps it speaks to the large target audience of youthful Android phone users or perhaps to the demographics of the people working for the app development companies, but the second most common class of apps was devoted to counting calories at fast food restaurants. Several applications count calories and nutrition at other types of restaurants, especially restaurant chains. These applications speak to the consistency of the meals offered in chain restaurants (whether the customer orders the chicken fajitas off the Chili's menu in Texas or California, the food is prepared in the same way and in close to the same size portion) so that the application has wide geographic applicability. The prevalence of these applications also speaks to the fact that chain restaurants now have a majority share of the restaurant marketplace. Four of the top 100 applications give calorie and nutrient information about foods in general (rather than about restaurant-prepared foods).

The second most common category of applications are those for locating nearby restaurants. Of these, the largest number deal with finding fast food restaurants. Some of these applications have geolocation information embedded or employ a map service such as Google Maps (e.g., the Fast Food app), and at least one touts its ability to find fast food restaurants faster than other apps (Find Food Fast). One app identifies fast food restaurants along an interstate highway (I-95 Exit Guide). Several help the user find food trucks, some of which change locations during the day. One app (Toptable Restaurant Finder) locates restaurants more generally – not only fast food restaurants – and includes the ability to book a reservation. Another app (Foodspotting) evaluates individual menu items at restaurants, since not every dish at a high-end restaurant is of high quality and even "dives" might offer one good item. Building on the popularity of food shows on television's Food Network, one app (Food Network On the Road) locates restaurants that have been featured on *Diners, Drive-Ins, and Dives* or other television programs.

Also popular among the Android food applications are those that help a user to find recipes. While there are two general recipe apps, most of them focus on a niche food market. One company produces the Food Street applications, a collection in which each individual app is about a particular kind of food such as vegetarian, diabetic, heart-healthy, low calorie, Puerto Rican, or Mexican. Other companies have created apps that cover other recipe food niches such as Chinese (Chef Panda), Thai (Thai Food Recipes), Indian (Sweet'N'Spicy), soul food (Soul Food Recipes), and party foods (Party Food Recipes).

It is notable how many food-related games show up as apps. There are food trivia games for adults (Food Trivia) and food memory games for children (Memory Game for Children – Fast Food). There are food shooting games akin to Angry Birds (Doodle Food Expedition) and fast food puzzles (Fast Food Mayhem). One can "drag and touch to move, eat and pop fish" in Fish Food. Some games involve operating or owning restaurants, such as Restaurant Live in which you and your friends open a five-star restaurant and have to attract customers; or Stand O' Food in which you "feed a host of hungry patrons in this fast-paced time-management challenge." (Quotations above are all from the app descriptions given in the Android Marketplace).

1.6 Conclusions

The initial section of this chapter provided an overview to the complicated and extensive issue of food online. The categorization offered there does a better job at describing the public face of individuals and companies as they interact with food topics online; a less good job at looking at the use of the Internet for business operations within a company or between a company and its suppliers. The four categories include one about individuals going about their private lives, one about individuals pursuing a professional life, and two about companies – divided into products and services. We have shown how to further subdivide these categories, depending on whether or not a company takes advantage of social networking on its website, whether it is a food-specific company or a more general company, and whether it has a physical presence or is online only.

The second section of this chapter explored the category 3 (corporate product) and category 4 (corporate service) websites. In particular, we have used Alexa data to provide information about the popularity of individual food websites compared with one another and compared against websites intended for other purposes; and we have given some groupings of the most popular food websites into categories.

The third section of the chapter provides further elaboration of category 1 (private individual) and category 2 (professional individual) websites. We have done this by considering three case studies. First is the story of Mrs. Q and her school lunch blog, which is the story of someone's online presence evolving from category 1 (a website intended for friends and family) to category 2 (a website for a public with a political purpose) and in the end, back again to a category 1 website. Second is the story of Kelly S. and the Yelp Elite Squad, which is the story of individuals using the medium of a category 4 website to build social capital for themselves. Finally, we relate the story of Joy the Baker, which details a particular kind of category 2 actor who becomes a category 3 actor by building a business out of a cooking hobby.

The last section complements the discussion in the earlier sections about websites by examining food applications for smartphones. It gives a categorization of the major food apps for the Android platform. It shows that all of these apps are for individual use to do things such as count calories, find restaurants, and play food-related games.

References

Balla L (2008) The week in Yelp: A day in the life of 'elite' Yelpers. Eater LA http://la.eater.com/archives/2008/07/24/the_week_in_yelp_a_day_in_the_life_of_the_elite_yelpers.php. 24 July 2008, Accessed 5 June 2013

Beck L (2013) This asshole is what happens when Yelp goes to your head. Jezebel. http://jezebel.com/5978190/this-asshole-is-what-happens-when-yelp-goes-to-your-head. 23 Jan 2013, Accessed 5 June 2013

Chen A (2011) Elite Yelp cult wreaks havoc on restaurants and hors d'oeuvres. Gawker. http://gawker.com/5808241/elite-yelp-cult-wreaks-havoc-on-restaurants-and-hors-doeuvres. 3 June 2011, Accessed 5 June 2013

Cortada J (2008) New wine in old and new bottles: Patterns and effects of the Internet on companies. In: Aspray W, Ceruzzi PE (eds) The Internet and American business. MIT Press, Cambridge, pp 329–422

D'Arcy J (2011) Fed up with school lunch, and with being anonymous. Washington Post. http://www.washingtonpost.com/blogs/on-parenting/post/sarah-wu-fed-up-with-school-lunch-and-with-being-anonymous/2011/10/18/gIQALuMKvL_blog.html 19 Oct 2011, Accessed 31 May 2013

Dell'Antonia KJ (2011) Double x book of the week: Fed up with lunch. Slate. http://www.slate.com/blogs/xx_factor/2011/10/07/fed_up_with_lunch_by_sarah_wu_is_the_doublex_book_of_the_week_.html. 7 Oct 2011, Accessed 31 May 2013

Eng M (2011) Tracking down Mrs. Q, AKA mom, blogger, and CPS employee Sarah Wu. Chicago Tribune. 5 Oct 2011

Forbes staff (2010) Eight of the very best food bloggers. Forbes. http://www.forbes.com/2010/03/03/best-blogger-food-lifestyle-forbes-woman-time-ree-drummond_slide_4.html. 3 March 2010, Accessed 1 June 2013

Los Angeles Times Staff (2011) A chat with Joy Wilson of Joy the Baker. Los Angeles Times. http://latimesblogs.latimes.com/dailydish/2011/10/a-chat-with-joy-wilson-from-joy-the-baker.html 7 Oct 2011, Accessed 1 June 2013

McHugh M (2013) What it's like being an outsider invited inside the Yelp elite circle. Digital Trends. http://www.digitaltrends.com/social-media/what-its-like-being-an-outside-at-a-yelp-elite-event/. 18 May 2013, Accessed 5 June 2013

Mrs. Q (2010a) My idea. Fed up with Lunch. http://fedupwithlunch.com/2010/01/eating-in-schools/. 3 Jan 2010, Accessed 22 June 2013

Mrs. Q (2010b) Day 80: Meatloaf. Fed up with Lunch. http://fedupwithlunch.com/2010/05/day-80-meatloaf/. 17 May 2010, Accessed 22 June 2013

Mrs. Q (2010c) Day 120: Hamburger, real potatoes! (and a bit about blogging). Fed up with Lunch. http://fedupwithlunch.com/2010/10/day-120-hamburger-real-potatoes-and-a-bit-about-blogging/. 7 Oct 2010, Accessed 22 June 2013

Mrs. Q (2010d) Day 133: Cheeseburger with a side of Twitter. Fed up with Lunch. http://fedup-withlunch.com/2010/10/day-133-cheeseburger-with-a-side-of-twitter/. 30 Oct 2010, Accessed 22 June 2013

Mullen M (2012) Joy Wilson of Joy the Baker. http://theeverygirl.com/feature/5285/. Accessed 1 June 2013

Nguyen T (2012) How elite Yelpers tried to extort Big Gay Ice Cream from their unopened store. The Braiser. http://www.thebraiser.com/big-gay-ice-cream-vs-yelp/. 27 Nov 2012, Accessed 5 June 2013

Owyang J (2008) Understanding community leadership: An interview with a member of Yelp's 'elite'. http://www.web-strategist.com/blog/2008/05/27/understanding-community-leadership-an-interview-with-a-member-of-yelps-elite/. 27 May 2008, Accessed 5 June 2013

Rachel, Recruiting (2011) Day in the life of a Yelp community manager. http://officialblog.yelp.com/2011/07/day-in-the-life-of-a-community-manager.html. 6 July 2011, Accessed 5 June 2013

Sax D (2011)Yelp's elite, epicurean force of totally free labor. Today News. http://www.today.com/id/43344769/ns/today-today_news/t/yelps-elite-epicurean-force-totally-free-labor/#.Ua9jBqWaErh. 12 June 2011, Accessed 5 June 2013

Searchquotes (2013) I want to be successful in everything I do, but the Internet and food just distract me. http://www.searchquotes.com/quotation/I_want_to_be_successful_in_everything_I_do,_but_the_Internet_and_food_just_distract_me./347388/. Accessed 22 June 2013

Stocker K (nd) Kelly Stocker. LinkedIn. http://www.linkedin.com/in/kellystocker. Accessed 5 June 2013

Susan, Yelp Recruiter (2009) A day in the life of a Yelp community manager. http://officialblog.yelp.com/2009/11/a-day-in-the-life-of-a-yelp-community-manager.html. 12 Nov 2009, Accessed 5 June 2013

Weiward Girl (2009) Confessions of a Yelper. http//weiwardgirl.wordpress.com/2009/02/22/confessions-of-a-yelper/. 22 Feb 2009, Accessed 5 June 2013

Vyvacious (2012) I'm Yelp elite, bitchess!!! :P. http://vyvacious.com/2012/11/20/yelp-elite-bitchess/. 20 Nov 2012, Accessed 5 June 2013

Zhao E (2011) Fed up with lunch blogger who exposed bad school lunches revealed as Sarah Wu on Good Morning America (video). Huffington Post. http://www.huffingtonpost.com/2011/10/05/fed-up-with-lunch-blogger_n_996972.html. 5 Dec 2011, Accessed 22 June 2013

Chapter 2
Anatomy of a Dot-Com Failure: The Case of Online Grocer Webvan

> *E-tailing does work. By using the Internet intelligently,*
> *old-economy grocers such as Safeway and Albertsons, far from*
> *being dinosaurs, are the wave of the future.*
>
> (Munroe 2013)

In 2008, when CNET published its list of the top ten dot-com flops ever, online grocer Webvan topped the list (German 2012). This chapter describes the rise and fall of Webvan, and analyzes the reasons for its failure. In particular, this case demonstrates that Internet companies – contrary to what many entrepreneurs believed during the dot-com boom – are not immune to the basic laws of economics or sound business practice.

2.1 Origins

In 1996, using funds from the sale 4 years earlier of the successful bookstore chain he had created with his brother, Louis Borders formed a new firm named Intelligent Systems for Retail. The computerized inventory system that the Borders brothers had built in order to customize the stock in each local Borders bookstore, which had been a major factor in the chain's success, had convinced Louis that the intelligent management of inventory and delivery offered many new business opportunities. Louis hoped that his new company could revolutionize direct delivery to consumers of a wide variety of products – offering personalization through the intensive use of technology.

Drawing on the cachet of his previous entrepreneurial success, Louis lined up a broad group of first-class investors – Benchmark Capital, Sequoia Capital, Yahoo, Softbank, Goldman Sachs, Barksdale Group, CBS, Knight Ridder, LVMH (Louis Vuitton, Moet, Hennessey), Amazon, and others – in addition to investing some of his own personal fortune. He attracted George Shaheen, the managing partner of

W. Aspray et al., *Food in the Internet Age*, SpringerBriefs in Food,
Health, and Nutrition, DOI 10.1007/978-3-319-01598-9_2,
© William Aspray, George Royer, Melissa G. Ocepek 2013

Andersen Consulting, to become the CEO. The company had a highly successful initial public offering, raising $375 million. Through these various efforts, the company acquired $1.2 billion in capital.

Borders's first venture partner, Benchmark, reined in his vision somewhat, and the decision was made to focus on selling groceries online, which would be delivered directly to the customer's home within a 30 min window of time specified by the customer on the previous day. The company was renamed Webvan Group in 1997 and became a leading player in the e-commerce field. Borders believed the company could be successful if it could capture even a small piece of the more than half trillion dollar American grocery market (Gale 2002; McAfee and Ashiya 2001).

The initial plan was for Webvan to open operations in 26 metropolitan areas. Customers would order online, with a toll-free telephone number available if they had questions. Orders would be filled from a high-tech, purpose-built warehouse located within the metropolitan area and delivered by one of the company's vans to the customer's home, where the groceries were left on the front steps or brought in to the kitchen. Webvan ordered untested high-tech warehouses for each of the metropolitan areas in which it planned to operate from the locally headquartered but nationally prominent construction and engineering company, Bechtel Corporation, at a cost of $35 million apiece. Thus the total price tag for these warehouses was a staggering 1 billion dollars. The warehouses were intended to be the most automated in the world, each one with more than four miles of conveyor belts carrying shopping baskets through the building. When a basket reached a "pod" (an order fulfillment station), an electronic order flashed a light and the stock shelves moved so that the human assembler was made aware of which grocery item to select and place in the cart, without ever having to move more than 19 ft. Trucks then took the cart to one of a dozen docking stations, where the groceries were loaded on a van for home delivery.

2.2 Operational Practices and Business Mistakes

The company began operations in ten cities by the end of 2000. The customer base continued to grow, and most customers were reasonably happy with the variety (e.g. 300 kinds of vegetables) and high quality of the products, as well as the generally on-time service. However, the customer growth was not rapid enough to secure a profit. The company expanded into new cities too quickly and spent too much money not only on the warehouses, but also on the vans to deliver the food and computer systems and proprietary software to run the operations. These capital expenses, together with salaries for 3,500 employees, meant the company was burning through $125 million per quarter.

The company made a number of basic business mistakes that savvy brick-and-mortar companies would likely have avoided (PC 2011). There was a belief among dot-com business entrepreneurs at the time that rapid growth was more important

- Some people enjoy the experience of grocery shopping.
- People commonly like to pick their own meat and produce.
- People get menu ideas by looking at store shelves.
- People are confused by a large number of provduct choices, which Webvan offered.
- People want to be able to use coupons, which Webvan allowed only near the end of the company's life.
- People want to be able to purchase economy-size budget packs of diapers and paper towels, which Webvan did not offer because these items take up too much space in the warehouse.
- People cannot easily commit in advance to a particular time to be home for the delivery.
- Women (the main target customers) were sometimes worried that they might be criticized for not doing their part for the household by making trips to the grocery store.
- People like to have the chance to make their own real-time decisions about substitutions if a product is not available on the shelf.
- A trip to the grocery is often coupled with other errands, such as going to the gas station or dry cleaners, so online grocery shopping does not eliminate household tasks outside the home.
- By not going to the store, people lose the opportunity to conduct other grocery store activities such as cashing a check, filling a prescription, or returning an item for immediate credit.
- People like to talk to store personnel and ask questions.
- People like to look at "meal-replacement" items such as ready-to-eat or ready-to-bake items before selecting them for purchase.

Fig. 2.1 Knowledge Webvan might have gained from user testing (Source: Hiser et al. 1999; Keh and Shieh 2001; Morganosky and Cude 2002; Ramus and Nielsen 2005)

than anything else; and many of these mistakes can be attributed to this "get big fast" philosophy. The 350,000 square foot warehouses, each of which could supply the equivalent of 18 grocery stores, were too large for the amount of business the company had attracted. In many cities, operations were running at only one-third capacity, well below the break-even point; and it was only through productivity gains in the warehouse that the company had any chance of competing with brick-and-mortar supermarkets. The design of the warehouses was not tested before Bechtel began to build them in multiple copies for the various cities in which Webvan was planning to operate; and a number of features had to be abandoned, such as the butcher area, which was closed when it was decided the company would do better to outsource meat provision, or when they learned from experience that the automated lazy susan slowed to a crawl in the deep freezer. The company apparently never conducted consumer testing, such as focus groups or surveys, to learn if there was sufficient demand for online grocery shopping. See Fig. 2.1 for a list of some of the things the company might have learned from user testing. It was also costing the company too much – $210 – to acquire each new customer. Many people tried Webvan once, but more than half of the first-time customers never returned.

2.3 Demise

It was probably a mistake for Webvan to have chosen groceries as its principal business, for the grocery business has one of the leanest profit margins (under 5 %) of any commercial market they could have entered. Moreover, none of the principals had experience in the supermarket business (Glasner 2001). Brick-and-mortar grocery chains had a competitive advantage against Webvan by filling online orders out of their stores without the time and capital expense of building a new warehouse infrastructure. The large brick-and-mortar supermarket chains had buying power that Webvan lacked. The company also made a mistake in selecting such a narrow delivery window. With only a small number of scattered customers, the 30 min window made the delivery portion of the operation highly inefficient and expensive to carry out. Peapod, one of Webvan's major competitors in the online grocery business, chose a 2 h window for delivery, which enabled it to rationalize its delivery routes to a far greater extent than Webvan.

Rather than trying to correct most of these problems, Webvan plowed ahead with its get big fast philosophy. In 2000, instead of cutting back and controlling its costs, the company expanded into Atlanta and Dallas. That same year, it bought its nearest rival in the online grocery business, Home Grocer, for $1.2 billion in stock. However, Webvan was never able to integrate the two lines of business sufficiently and retain the customers so as to gain advantage from the merger (Cuneo 2000a, b).

In one last-ditch effort, after the company had started to hemorrhage money, it decided to rebrand itself as a purveyor not only of food, but also of electronics, pet supplies, kids clothing, non-prescription pharmaceuticals, books, and a half dozen other product categories. This compounded the problem because the company's original brand and mission were not yet well known and rebranding was expensive.

In July 2001 Webvan laid off its remaining 2,200 employees and ceased operations in the seven markets in which it was then operating (San Francisco, Los Angeles, Orange County, San Diego, Portland, Seattle, and Chicago), having gone through $1.2 billion in only 2 years of set-up and another 2 years of operation (CNN Money 2001; Delgado 2001; Hansell 2001a, b; Knowledge@Wharton 2001; Levis 2011; Venture Navigator 2007; Weiss 2001). At the end, the company had approximately 750,000 customers – although only a small percentage of them used the service regularly. A month later, stock certificates from Webvan were being sold on eBay as memorabilia of "the most spectacular failure in Internet business history" (eBay sales copy, as quoted in Tedeschi 2001).

2.4 Competitors in the Online Grocery Business

Stepping away from the individual story of Webvan and placing it in a larger business context, the first online grocer was Grocery Express, which began operations in San Francisco in 1981. In the mid and late 1980s, Grocery Express customers

1. Marsh
2. Harris Teeter
3. Giant Food
4. Albertsons
5. Stop & Shop
6. Bashas
7. Peapod
8. Schnucks
9. Hy-Vee
10. Ingles Markets
11. Why Run Out
12. Stater Bros.
13. Easy Grocer
14. Publix
15. Walgreens
16. Simon Delivers
17. My Web Grocer
18. Market One Stop
19. House Calls Online
20. Price Chopper
21. Net Grocer
22. Ethnic Grocer
23. Grocer Online
24. Kroger
25. Metro Food Market
26. Electric Food
27. Your Grocer
28. Groceries Express
29. Bluelight
30. Giant

Fig. 2.2 Top-ranked online grocers as of third quarter 2001 (Source: Lim et al. (2004) [MyWebGrocer is a company that provided hardware and software so that brick-and-mortar groceries such as Piggly Wiggly and D'Agostino's could add an online operation])

connected through the Prodigy online service, before the World Wide Web existed. Among the earliest entrants, in addition to Webvan and Grocery Express, were Streamline, Shoplink, and Homeruns (Weise 1999). All of these companies were "pure" players, i.e. they started as Internet companies rather than being brick-and-mortar groceries that later entered the online grocery business. Streamline and Shoplink took a different approach from Webvan in that they used reception boxes rather than requiring the customer to be home for the delivery. Reception boxes are locked, refrigerator-like boxes with separate compartments for frozen, refrigerated, and room-temperature groceries. Reception boxes ease the logistics of delivery and free customers from having to be home when the delivery is made (Ring and Tigert 2001; Smaros and Holmstrom 2000; Tanskanen et al. 2002; Williamson 2000; Willoughby and Holcomb 2001; Yrjola 2003). However, neither of these companies was successful. Figure 2.2 shows the list of U.S. online grocers as of the third quarter of 2001 (ranked according to an analysis that shows various qualities, not amount of business). Note that some of the leaders at that time were pure players. By the first quarter of 2004, however, none of the pure players ranked in the top ten. The companies that were most successful in the online grocery business were the brick-and-mortar grocery firms that used online sales to extend their services and capitalize on their existing infrastructures. Change occurred quickly in this field. Of the 30 companies listed as players in the first quarter of 2001, ten were no longer in the business by the fourth quarter of 2002. Note that two of the brick-and-mortar companies in online groceries were not even grocery firms: Walgreens is a drug store chain and Bluelight is the online website for discount retailer Kmart. The two largest brick-and-mortar grocery firms in the United States, Kroger and Walmart, were hardly a presence in the online grocery business at this time.

The year 2000 was a principal entry year into the online grocery field (Streif 2000). Two of WebVan's major startup competitors were Kozmo and UrbanFetch – both founded that year (Moskin 2005). 28-year-old New York City former investment banker Joseph Park founded Kozmo as an online convenience store (Scott 2002). The customer who did not want to make the trek to the corner store could order soft drinks, ice cream, movie rentals, and other convenience store items online, and they would be delivered by bicycle courier to the home within an hour (Collins 1999; Kushner 1998). UrbanFetch had a similar business model and also delivered food, books, and compact discs by bicycle in Manhattan and London, but most of its revenue came from delivering – within an hour – higher-priced items such as designer fragrances, video games, digital cameras, and DVD players.

Kozmo started its business operations in New York City with many of its early employees sleeping in the company's offices and showering at local health clubs. Amazon pledged a $60 million investment in Kozmo, when it agreed to deliver books, music, and toys for Amazon by bicycle courier within an hour of being ordered (in the cities in which Kozmo operated) (Dobbins 2000). Kozmo also entered into a 5-year agreement to place dropboxes for its movie rentals in Starbucks coffee shops and deliver Starbuck coffee products. This arrangement with Starbucks did not work out well; it cost Kozmo $150 million and was terminated after only 13 months. Kozmo originally delivered products for free, but eventually it instituted a $4.95 delivery fee. Even that fee did not cover the actual cost of delivery (estimated at $10), nor was there enough of a margin in the markup on video rentals or pints of ice cream to subsidize the delivery cost. Kozmo considered acquiring UrbanFetch in 2000 until it learned more about the latter's financial circumstances (Fendelman 2000). UrbanFetch closed in 2000, followed by Kozmo in 2001. Streamline and Shoplink also soon closed.

While Webvan was not able to succeed in the online grocery market, other organizations have (Munroe 2001). In Britain, the Tesco supermarket chain has had a profitable home grocery delivery service since 1996 (Delaney-Klinger et al. 2003; Ehrlich 2006). Tesco decided to fulfill its delivery orders from within its existing brick-and-mortar stores, using carts that follow a computer-generated path through the store, filling six orders simultaneously, and employing efficient delivery routes distributed across the metropolitan area. Tesco charges an upfront fee of five pounds for the service. Its success in this business in the United Kingdom led it to operate a successful grocery delivery service in the United States for the Safeway grocery chain.

Some startups have also succeeded, notable among them is Peapod (now part of the Dutch Royal Ahold grocery conglomerate). Today, Peapod operates a profitable grocery delivery service in connection with two brick-and-mortar grocery chains also owned by Royal Ahold: Giant Food in Pennsylvania, Maryland, Virginia, and Washington, DC; and Stop & Shop in Connecticut, Massachusetts, New Jersey, and New York. Peapod has succeeded by doing a number of things differently from Webvan. Peapod does not open a centralized warehouse until the volume of business makes this a cost-effective practice, and in smaller markets it fulfills orders from small buildings adjacent to existing grocery stores. In the beginning, Peapod's fulfillment was carried out through existing brick-and-mortar grocery stores. Peapod tested markets before making a large commitment – in contrast to Webvan's strategy of getting big fast. Peapod had a 2 h time delivery window, making it easier to rationalize delivery routes and reduce delivery costs. Peapod focused more on business practices than on technology, and it did not incur the large up-front costs for technological infrastructure that burdened Webvan. Peapod's management was already familiar with the grocery business and was careful to learn lessons from its interactions with customers (Lunce et al. 2006).

Others successful in the online grocery business in the United States include Coborn Delivers (formerly Simon Delivers) in Minnesota and Wisconsin, FreshDirect in New York City, and Winder Farms in the western United States. The brick-and-mortar grocery chains Albertson's, HyVee, and ShopRite have also operated online grocery delivery services. Amazon, which was an investor in Webvan, has operated its AmazonFresh delivery system in Seattle since 2007. It can be accessed through all of the major mobile phone platforms. Even the hotel and airline ticket discounter Priceline joined the business with its WebHouse Club, which for a $3 monthly membership fee allowed customers to set the price at which they were willing to pay for groceries; and if a grocer accepted their bid, the customer would pick the groceries up at a Giant or Safeway store (Canedy 1999).

2.5 Business Challenges and Opportunities for Online Grocers

Turning away from the story of individual firms such as WebVan or Kozmo, what does the academic literature tell us about online grocery shopping? Online grocery shoppers are typically young (less than 45 years of age), female, college graduates, and with a household income over $70,000. Older people, major shoppers, and people with lower income are less likely to know about online grocers; and older people and those with no college education are less likely to consider using them. Perception of convenience and the ability to save time are the strongest drivers of online grocery shopping. Consumer trust is higher in firms that have a physical as well as an online presence. Individuals who have experienced online grocery shopping are generally more trusting of it, and are even willing to buy produce and meat online. Perceived advantages of online groceries are greater convenience, wider product offerings, and better prices; disadvantages include not being able to check the products (fear of inferior quality) and loss of recreational shopping (Hansen et al. 2006).

One of the major business concerns in online groceries is the last-mile delivery problem. Compared to Webvan's business practice of attended delivery – even compared with a more flexible 2 h delivery period rather than Webvan's 30 min delivery window – having a reception box at the customer's house lowers delivery transportation costs by 43–53 % because the delivery routes can be rationalized more effectively. If the online grocer sets up shared reception boxes, perhaps at the local convenience store or gas station, the delivery cost is reduced further – 55–66 % lower than home-attended delivery. However, the use of reception boxes introduces capital expense for the construction and delivery of the boxes, which has to be recovered over time. One study argued that it would be impossible to get sufficient density for an attended reception model to work generally in the United States (Cox 2011; Kamarainen 2003; Kamarainen et al. 2001; Punakivi 2003).

The academic literature also has considered grocery distribution facilities. Use of a specialized distribution center, like those Webvan built, can be more efficient than using a local grocery store – but only in cases of high volumes of business. In other cases, using the grocery store infrastructure is clearly more efficient and has the added benefit of fast startup and low initial investment. Automation of the grocery picking (as Webvan used) only works in situations where there is high and constant demand; in many cases, these studies found, a flexible distribution with manual picking is more effective. The studies also argue that online grocery success depends on local factors, and thus it is better to do well in a confined geographical area before opening in another area – contrary to Webvan's plan to rapidly enter multiple cities.

Finally, the literature notes how formidable a competitor the traditional brick-and-mortar grocery is. Quoting one study (Ring and Tigert 2001, p. 271):

> Supermarket chains in most cities have developed a loyal customer base through years of service and convenient locations. Supermarkets are extremely efficient, with total cost per customer served of approximately $6-7. They are constantly advertising to maintain high levels of awareness about prices, assortments, service, quality, specialty products and loyalty programs.

Clearly, Webvan did not understand the business it was getting in to. There seems to be some limited promise for online groceries, but it is an open question whether they are viable in geographic regions with low-density populations.

2.6 Conclusions

Webvan was a monumental failure. It proved that Internet companies are not immune from the basic laws of economics and the sound practices of business. In particular, Webvan serves as a poster child for the problems of excessive spending and the get-big-fast philosophy characteristic of the dot-com boom and bust. It was hubris on the part of the entrepreneurs who established Webvan to rely on their native smarts and experience and spent more than a billion dollars without first gaining a better understanding of its customers' needs and wants. The company did not realize how many grocery shoppers enjoy the experience of shopping, want to select their own meat and produce, are overwhelmed by too many product choices, and find it problematic to schedule in advance to be home for a delivery (even if the delivery window is small). Subsequent developments have shown that a small market for online grocery shopping and delivery does exist, provided that capital outlay is minimized and infrastructure is piggybacked on that of brick-and-mortar grocery stores. Webvan also made the mistake of testing its service in a market close to home in Silicon Valley rather than in a community where density of housing (e.g. Manhattan) or absence of brick-and-mortar groceries (e.g. Detroit) would have made for better prospects of success. The Internet parts of this business – ordering and payment – seem to be the least important elements for success. The logistics and warehousing, as well as the analytics of what products to offer and how to design the delivery routes and timetables, are the critical issues in making this business successful.

References

Canedy D (1999) What's your bid on peanut butter? New York Times. http://www.nytimes.com/1999/09/22/nyregion/what-s-your-bid-peanut-butter-groceries-join-big-items-name-your-price-web-site.html. 22 Sept 1999, Accessed 6 June 2013

CNN Money (2001) Webvan shuts down. http://money.cnn.com/2001/07/09/technology/webvan. 9 July 2001, Accessed 6 June 2013

Collins G (1999) Selling online, delivering on bikes. New York Times. http://www.nytimes.com/1999/12/24/nyregion/selling-online-delivering-bikes-low-tech-courier-services-thrive-growth-web.html?pagewanted=all&src=pm. 24 Dec 1999, Accessed 6 June 2013

Cox N (2011) Acceptance of e-grocery: An empirical study about the use and potential use of online grocery shopping. Faculty of Economics and Business Administration, Maastricht University, Maastricht, Netherlands

Cuneo AZ (2000a) Peas fill up the pod; E-grocers win with customers, lose on Wall Street. Advertising Age. p 52, 3 April 2000

Cuneo AZ (2000b) Webvan campaign plays upon anti-grocery store sentiment. Advertising Age. p 58, 17 April 2000

Delaney-Klinger K, Boyer KK, Frohlich M (2003) The return of online grocery shopping: A comparative analysis of Webvan and Tesco's operational methods. TQM Magazine 15(3): 187–196

Delgado R (2001) Webvan goes under. San Francisco Chronicle. http://www.sfgate.com/cgi-bin/article.cgi?f=/c/a/2001/07/09/MN196371.DTL&ao=all. 9 July 2001, Accessed 6 June 2013

Dobbins A (2000) Amazon puts $60 million in 1-hour e-delivery system. Newsbytes. http://betanews.com/2000/03/20/amazon-puts-60-million-in-1-hour-e-delivery-system. 20 March 2000, Accessed March 13, 2012

Ehrlich P (2006) Webvan vs. Tesco groceries. http://www.math.ufl.edu/~ehrlich/webvan.html. 3 Jan 2006, Accessed 8 Feb 2012

Fendelman A (2000) Kozmo calls off Urbanfetch acquisition. ePrairie.com. http://www.lexisnexis.
    com.ezproxy.lib.utexas.edu/lnacui2api/api/version1/getDocCui?lni=41DB-NX10-00G8-P4P4
    &csi=167445&hl=t&hv=t&hnsd=f&hns=t&hgn=t&oc=00240&perma=true. 11 Oct 2000,
    Accessed 6 June 2013
Gale Encyclopedia of E-Commerce (2002) Webvan Group, Inc. http://ecommerce.hostip.info/
    pages/1080/Webvan-Group-Inc.html. Accessed 7 Feb 2012
German K (2012) Top 10 dot-com flops. CNET. http://www.cnet.com/1990-11136_1-6278387-1.
    html. Accessed 8 Feb 2012
Glasner J (2001) Why Webvan drove off a cliff. Wired. http://www.wired.com/techbiz/media/
    news/2001/07/45098. 10 July 2001, Accessed 6 June 2013
Hansell S (2001a) Some hard lessons for online grocer. New York Times. http://www.nytimes.
    com/2001/02/19/business/some-hard-lessons-for-online-grocer.html?pagewanted=
    all&src=pm. 19 Feb 2001, Accessed 6 June 2013
Hansell S (2001b) An ambitious Internet grocer is out of both cash and ideas. New York Times.
    http://www.nytimes.com/2001/07/10/business/an-ambitious-internet-grocer-is-out-of-both-
    cash-and-ideas.html?pagewanted=all&src=pm. 10 July 2001, Accessed 6 June 2013
Hansen T, Solgaard HS, Cumberland F (2006) Determinants of consumers' adoption of online
    grocery shopping. Eur Adv in Consum Res 7:276–277
Hiser J, Nagaya RM, Capps O Jr (1999) An exploratory analysis of familiarity and willingness to
    use online food shopping services in a local area of Texas. J of Food Distrib Res 30(1):78–90
Kamarainen V (2003) The impacts of investments on e-grocery logistics operations. Dissertation,
    Helsinki University of Technology.
Kamarainen V, Saranen J, Holmstrom J (2001) The reception box impact on home delivery effi-
    ciency in the e-grocery business. Int J of Phys Distrib and Logist Manag 31(6):414–426
Keh HT, Shieh E (2001) Online grocery retailing: success factors and potential pitfalls. Bus Horiz
    44(4):73–83
Knowledge@Wharton (2001) Webvan finds that shopping for food online hasn't clicked with cus-
    tomers. http://knowledge.Wharton.upenn.edu/article.cfm?articleid=321. 19 March 2001,
    Accessed 6 June 2013
Kushner, D (1998) Going out, without leaving home. New York Times. http://www.nytimes.
    com/1998/06/11/technology/going-out-without-leaving-home.html?pagewanted=all. 11 June
    1998, Accessed 7 June 2013
Levis K (2011) Webvan. www.kieranlevis.com/workshop-for-imperial-mba-students-3-march-
    2011/webvan. Accessed 7 Feb 2012
Lim H, Heilig JK, Ernst S, Widdows R, Hooker NH (2004) Tracking the evolution of e-grocers:
    A quantitative assessment. J of Food Distrib Res 35(2):66–76
Lunce SE, Lunce LM, Maniam B (2006) Success and failure of pure-play organizations: Webvan
    versus Peapod, a comparative analysis. Ind Manag and Data Syst 106(9):1344–1358
McAfee A, and Ashiya M (2001) Webvan. Case, Harvard Business School. http://hbr.org/product/
    webvan/an/602037-PDF-ENG. 25 Sept 2001, Accessed 6 June 2013
Morganosky MA, Cude BJ (2002) Consumer demand for online food retailing: Is it really a supply
    side issue? Int J of Retail and Distrib Manag 30(10):451–458
Moskin J (2005) Online shopping makes New York a cardboard jungle. New York Times. http://
    www.nytimes.com/2005/04/06/dining/06fres.html. 6 April 2005, Accessed 6 June 2013
Munroe T (2001) Intelligent e-tailing: avoiding Webvan's mistakes. Turnaround Management
    Association. http://www.turnaround.org/Publications/Articles.aspx?objectID=1802. 1 Aug
    2001, Accessed 6 June 2013
PC (2011) Why did Webvan fail so spectacularly? The Idea Post (blog). http://theideapost.blogspot.
    com/2011/02/why-did-webvan-fail-so-spectacularly.html. 1 Feb 2011, Accessed 6 June 2013
Punakivi M (2003) Comparing alternative home delivery models for e-grocery business.
    Dissertation, Helsinki University of Technology.
Ramus K, Nielsen NA (2005) Online grocery retailing: What do consumers think? Internet Res
    15(3):335–352

Ring LJ, Tigert DJ (2001) Viewpoint: The decline and fall of Internet grocery retailers. Int J of Retail and Distrib Manag 29(6):264–271

Scott AO (2002) Film review; Chronicling a bubble called kozmo.com. New York Times. http://www.nytimes.com/2002/01/11/movies/film-review-chronicling-a-bubble-called-kozmocom.html. 11 Jan 2002, Accessed 6 June 2013

Smaros J, Holmstrom J (2000) Reaching the consumer through e-grocery. Int J of Retail and Distrib Manag 28(2):55–61

Streif T (2000) Web-based companies making urban delivery inroads. Deutsche Presse-Agentur. 21 Feb 2000

Tanskanen K, Yrjola H, Holmstrom J (2002) The way to profitable Internet grocery retailing: Six lessons learned. Int J of Retail and Distrib Manag 30(4):169–178

Tedeschi B (2001) The fallen dot-coms are not yet cold, but some sealers are already selling their detritus as memorabilia. New York Times. http://www.nytimes.com/2001/08/13/business/e-commerce-report-fallen-dot-coms-are-not-yet-cold-but-some-dealers-are-already.html?pagewanted=all&src=pm. 13 Aug 2001, Accessed 6 June 2013

Venture Navigator (2007) Webvan's unsustainable business model. http://www.venturenavigator.co.uk/content/153. Aug 2007, Accessed 6 June 2013

Weise E (1999) Delivering what you want now. USA Today. 14 Dec 1999

Weiss TR (2001) Online grocer Webvan crashes with a thud. Computerworld. http://www.computerworld.com/s/article/62171/Online_Grocer_Webvan_Crashes_With_a_Thud. 16 July 2001, Accessed 6 June 2013

Williamson DA (2000) Why net delivery service schemes are out of order. Advertising Age. http://adage.com/article/news/net-delivery-service-schemes-order/1337. 21 Aug 2000, Accessed 6 June 2013

Willoughby C, Holcomb R (2001) Ready, get set, click: Grocery shopping online. In: food technology fact sheet. Stillwater: Robert M. Kerr Food and Agricultural Products Center, Oklahoma State University. http://pods.dasnr.okstate.edu/docushare/dsweb/Get/Document-980/FAPC-114web.pdf. FAPC-114, April 2001, Accessed 6 June 2013

Yrjola H (2003) Supply chain considerations for electronic grocery shopping. Dissertation, Helsinki University of Technology

Chapter 3
The Dark Side of Online Food Businesses: Harms to Consumers and Main Street Businesses

One of the biggest sources of indigestion for restaurant owners has been Yelp.

(Rowe 2009)

When faced with the extraordinary successes of Internet companies such as Amazon and Google and the advantages they provide to both their customers and affiliated businesses, it is perhaps difficult to see the harm that Internet-based companies can create. This chapter is intended to temper the sentiment about the virtues of the Internet. It examines three popular Internet-based, food-related companies and discusses how each has caused harm to consumers and Main Street businesses. More specifically, the chapter studies the restaurant reservations company OpenTable and how consumer expectations that restaurants will add online reservation systems burden restaurant owners with high costs and may harm the ambience of restaurants; the online coupon company Groupon and how businesses can be harmed by spike in demand and changed expectations of consumers; and the restaurant reviewing company Yelp and how it can ruin the surprise factor for new patrons, lead to hostilities between owners and reviewers, and harm restaurants through unsavory sales practices.

3.1 Case 1: OpenTable

Our first case study is of the restaurant reservation company OpenTable. It is among the ten most heavily trafficked food businesses online (ranked 623rd in US website traffic by Alexa.com in May 2013). More than 20,000 restaurants participate in the OpenTable reservation system, and more than a million reservations are made through it each month. Because the authors of this book, together with the business historian James Cortada, have published a study of this company elsewhere, we will only summarize the history and harms of OpenTable here (Aspray et al. 2013).

W. Aspray et al., *Food in the Internet Age*, SpringerBriefs in Food,
Health, and Nutrition, DOI 10.1007/978-3-319-01598-9_3,
© William Aspray, George Royer, Melissa G. Ocepek 2013

Restaurant reservation systems were among the earliest food applications to appear on the Internet, with 20 companies offering this service by 2000 (Halm 2000). However, it was not until OpenTable came along that any online service took an appreciable share of the reservation service business; prior to that, most reservations were made by telephone (On competition, see Foodie Buddha 2011; Fund Manager 2011; Livebookings 2011). OpenTable was founded in 1998 in San Francisco by Chuck Templeton, an entrepreneur who had worked his way through college holding various jobs in restaurants (Morell 2011). He left the company in 2004, and it has been professionally managed since then, funded by several rounds of public offerings, and expanded into a number of new metropolitan areas and various smartphone platforms (Austin 2009; Bloomberg Businessweek 2012; Davis 2012; Geron 2009; Hafner 2007; Joseph 2006; Lieberman 1999; Localized USA 2012; McLaughlin 2010; Primack 2011; Silicon Valley/San Jose Business 2008; Tedeschi 2004).

OpenTable is best known for its OpenTable Connect technology, which enables a user on a computer or mobile phone to specify a time and place when he wants to eat, then the Connect technology will present a list of all the restaurants enrolled with OpenTable that have an open table at or around the specified time, and a single click enables the customer to make a reservation. OpenTable also builds and installs Electronic Reservation Books, which computerize all of the restaurants host stand operations – not only reservations but also activities such as email marketing and business analytics.

To use OpenTable's services, restaurants are charged a one-time installation fee, a monthly licensing fee for use of the terminal and software, and a charge for each person who makes a reservation online – whether the reservation is made through OpenTable or the restaurant's own website. Some restaurant owners are unhappy with the cost of OpenTable. They feel a competitive pressure to offer the convenience of OpenTable reservations, even if they believe that the people who reserve on OpenTable would have reserved in another way if OpenTable had not been available, or that they could have filled all their tables without OpenTable. Restaurant owners are particularly unhappy about having to pay OpenTable for reservations when they are made through their own websites rather than through OpenTable. The costs mount up. One restaurant owner complained that the restaurant paid more to OpenTable than to its main chef, and others talked about how the fees paid to OpenTable cut into their advertising budgets (Lang 2006; Leung 2011; Restaurant Hospitality 2010).

Restaurant owners also complain about the impact of OpenTable on on-site operations. It takes a week of training for a restaurant employee to become skilled in using OpenTable's system. High-end restaurants are often concerned about giving the personal touch to their customers, and for some restaurants it means hiring an additional person for the host stand to meet and greet customers, while another host is busy with the computer system, managing reservations and tables. This kind of computerized system might be regarded as a back-office operation in many industries, but it is located up front at the host stand in restaurants. If the restaurant owner is concerned about the ambience in the restaurant, the OpenTable terminal may have to be screened off from the seating area, causing extra construction costs and perhaps reducing floor space for dining tables.

While individual diners generally like the convenience of OpenTable, it also has its drawbacks for them (Kimes 2009; Leson 2005; Sonnenschein 2011). Some negative comments about OpenTable have appeared in the online food discussions on Chowhound, where people argue that they received poorer service when they make a reservation through OpenTable, presumably because of the disgruntlement of the restaurant management with having to use this reservation software. Other individuals have raised the privacy concern that the restaurant enters their names into the Electronic Reservation Book and can track them, even if they were walkups or made the reservation by telephone.

3.2   Case 2: Groupon

Another type of business that has appeared online and accelerated with the wide distribution of personal computers and smart phones is e-coupons, sometimes known as daily deal coupons. This section profiles the market leader, Groupon, which has a company history that is not dissimilar from that of WebVan (described in Chap. 2) in experiencing the rewards and risks associated with a get-big-fast philosophy. Groupon has the 85th most web traffic in the United States, making it the second most trafficked food website in America (Alexa.com, as of May 2013).

Coupons are an old business practice, but they have received a new life in the age of the Internet and smart phones. Apparently, Coca-Cola offered the first business coupon in America, in 1888. By 1913, more than eight million consumers had tried a free glass of the "real thing," and the coupon industry was born with newspaper and magazine coupons, coupon books, and the like (Bowman 1985).

People sign up, at no cost, with one of the e-coupon companies such as Groupon, LivingSocial, or Google Offers. Each day, the Groupon member is given the opportunity to purchase at a reduced price from a small collection of products or services being offered in her local area. The consumer pays Groupon and then receives a coupon to use, for example at a local restaurant or a local spa. Groupon typically retains about half the revenue from the coupon purchase and pays the remaining funds to the restaurant. To make the couponing effort a worthwhile advertising venture for the restaurant, the deal does not consummate until a pre-determined number of customers have agreed to buy coupons. Once this tipping point is reached, the customers have their credit cards charged and they receive their coupons online (Cohen 2009; Restaurant Hospitality 2011b; Sennett 2012).

The company primarily responsible for building this model is Groupon, although today it has hundreds of competitors. Groupon's only real predecessor was Mercata, a start-up business founded by Microsoft co-founder Paul Allen in 1999 and closed in 2010, which allowed people to get discounts on items by banding together and buying them in bulk (Burkeman 2011). The founder (and CEO through early 2013) of Groupon was Andrew Mason (Coburn 2010; Williams 2010). Both of Mason's parents ran their own small businesses, and as a teenager he resold candy bought from Costco in the school cafeteria to generate spending money. His main interest was in music. After graduating with a music degree from Northwestern University in 2003, he went to work designing web pages for companies in the Chicago area. In Fall 2006 he quit his job to enter a master's degree in political science at the University of Chicago.

According to the creation myth, in 2006 Mason had difficulty resolving a contractual problem with his cell phone provider, and he began to think about how he could leverage more power in this kind of bargaining situation by uniting with other consumers. This led him to consider how to use social networking technology to organize people for collective actions that were too large for individuals to solve on their own – not only in this kind of business transaction but also, for example, in taking political action.

Mason worked as a tech employee for two companies, Echo Global and InnerWorkings, founded by Chicago venture capitalist Eric Lefkosky. One day in 2007 Lefkosky surprised Mason by offering him 1 million dollars to build the social networking platform Mason had been talking about. Mason dropped out of graduate school and developed ThePoint, which was a web-based technology that enabled people to sign up for a political or social cause but would not engage them until a pre-determined tipping point had been reached – a tipping point high enough that a collective action would have a reasonable chance to achieve the goals of the cause. Some of the uses of ThePoint in its first year were fanciful, such as an effort to sign people up to pay for a dome to be built over Chicago to keep out the harsh winter weather (This particular action never came close to reaching its tipping point, although it did attract some people who pledged $10,000!). About a year into the project, Lefkosky became uneasy about his investment and pushed Mason to move quickly to find an application of ThePoint that would yield a profitable return. Mason and Lefkosky agreed on collective buying, which quickly turned into Groupon, launched in November 2008.

The company initially set up business operations in Chicago and maintained its headquarters there. It grew at a breakneck pace, opening in a new city every few weeks (Steiner 2010). By 2010 it had enrolled over 70 million members, received more than $700 million in revenue, operated in almost 100 US cities and 25 other countries, and employed almost 10,000 workers (O'Dell 2010). It expanded into Europe by acquiring CityDeal, a German knock-off of Groupon that had unusually effective operations. When one thinks of an Internet company, one thinks of hordes of programmers. The technology at Groupon is simple, and the company employs mostly writers and sales people, not programmers. Many of the copywriters hired have experience as journalists. Employees hone their ability to write the witty, sometimes biting copy that Groupon has come to be known for in training sessions at what is known as the Groupon Academy (Weingarten 2010). An example of this style: "Skydiving is the perfect way to celebrate a birthday, sweat out premarital terror for a bachelor or bachelorette party, or take a glorious leap into a new life as a migratory swallow" (Groupon ad as quoted in Weiss 2010). A number of the early customer-service employees were actors or stand-up comics from Chicago's improv comedy scene. Most of the employees are young and the majority are women, perhaps because it is believed they will respond better to the large number of coupon offers directed primarily at women, e.g. for manicures or spa treatments. Mason attributes the success of Groupon to its ability to serve as a "discovery engine" by which people can learn more about the complex metropolitan area in which they live – e.g., try a new restaurant or do something such as rock wall climbing that they would not otherwise try. It is common for people to use Groupons to create social events with family or friends, e.g. go to a new restaurant together.

What Groupon is really offering to the restaurants is a marketing service. For small businesses, it can be a good alternative to traditional marketing. The small business receives exposure through the Groupon offer to every individual signed up with Groupon in that geographic region, whether or not the member elects to purchase the coupon to eat at the restaurant. The small business does not have to pay any money up front to Groupon as it would to a traditional marketing company. Instead, it pays for its marketing by receiving lower than normal revenue on the featured product it provides to customers. If the Groupon purchases $40 worth of meals at the restaurant for a price of $20, and the restaurant receives from Groupon half of the cost of the coupon, the restaurant receives only $10 in revenue to provide $40 worth of meals. (The restaurant often also receives additional revenue from the Groupon customer, at regular prices, for items purchased in addition to those covered by the coupon.) Using Groupon, the restaurant only needs to provide discounts to customers if it is assured of receiving enough new traffic (by having enough people purchase this coupon so that the tipping point is achieved).

According to a study by Rice University professor Utpal Dholakia (2011), covering the period August 2009 through March 2011, restaurants do worse economically than other types of companies that sign up for daily deal promotions with Groupon; 43.6 % of the restaurants earn a profit from their e-coupon deal, and only 35.9 % are willing to run another e-coupon promotion. Thus in the majority of cases, the restaurant loses money on each customer. A serious problem can arise when a large number of subscribers sign up for the coupon, since the per-coupon loss is multiplied many times over. From its earliest days, Groupon established a minimum number (the tipping point) at which the coupon went into effect. Not long into its history, however, Groupon decided to allow restaurants (and other businesses) to also set a maximum number of coupons that could be sold – as a means to limit the loss through these loss-leader promotions. Unfortunately, restaurants and other businesses that sign up with Groupon do not always understand the risk fully and might not be able to predict accurately the interest the coupon will attract to their product, so sometimes they do not set maximum coupon sales or set them too high (Streitfeld 2011b).

Two well-circulated examples show how damaging an overly successful Groupon subscription can be for a small business (Mui 2010). In one case, a British woman who had been making and decorating cupcakes for 25 years ran a Groupon promotion without setting a cap. She received Groupon orders from 8,500 customers, who ordered more than 100,000 cupcakes, and the losses on this promotion wiped out her entire year's profits (Wilkes 2011). In another case, the Groupon promotion for

Posie's Café in Portland, Oregon, which sold almost a thousand coupons, swamped the café with customers for 3 months – driving away some regular customers – and the owner had to take $8,000 out of his personal savings to pay for the extra labor needed to serve these new customers (BBC 2012; Carrera 2011).

Groupon attracts restaurants and other businesses to sign up for coupon deals through paid advertisements on Google and Facebook, and by intensive sales calls in locations where the company operates or wants to operate. The response has been strong; in fact, in its early history Groupon turned down seven out of eight companies that wanted to offer coupons so as to keep its product offerings of high quality. The biggest response to date to a Groupon offer is a $25 coupon for $50 worth of merchandise at The Gap clothing store; more than 400,000 coupons were sold. Groupon works at providing strong customer service, operating an around-the-clock hotline and following a no-questions-asked return of coupons policy. As of late 2010, 95 % of Groupon offerings were reaching their tipping point. Groupon improved its coupon marketing by personalizing the offers it sends to members, based on gender, location, and buying history. Groupon also made changes so that individual users can set personal preferences to only receive information about specific types of deals. The company has recently expanded its offerings with a location-based service oriented toward mobile phones called Groupon Now, which involves making offers in a small geographic area available only for a short duration (typically a few hours). It has also started several other businesses: Groupon Live, a partnership with Live Nation Entertainment to sell discounted tickets to live concerts; Groupon Getaways, to sell discounted travel and hotel deals in collaboration with Expedia; Groupon Goods, a direct seller of discounted goods; and Groupon Reserve, a premium service with exclusive offers for an elite audience.

Growing as rapidly as the company has, and offering service in so many locations, has required substantial capital. In addition to the initial funding received from Lefkosky, Groupon received $135 million from Battery Ventures in Silicon Valley, Digital Sky Technologies in Russia, and several other venture capital firms. Groupon turned down an offer in 2010 from Yahoo to purchase the company for an estimated $2–3 billion (Arrington 2010). Groupon surprised the investor community later that same year when it declined a takeover offer of $6 billion from Google – what would have been the largest acquisition ever made by Google (MacMillan and Galante 2010; Weiss 2010). Mason claimed the reason for this decision was his concern that the acquisition would lead to lower employee morale and worsened relations with business clients. Some analysts argue that it was instead because several Groupon board members believed that the Google offer undervalued the company, while other board members were concerned about attracting antitrust scrutiny. In 2011, Groupon had an initial public offering, which raised an additional $700 million.

Groupon faces a number of challenges from competitors (Perez 2011). The barriers to enter this business are low, and Groupon has not really found any scale effects that would give it a competitive advantage based on its large size. In fact, there are estimated to be 500 direct competitors to Groupon in the e-coupon industry. When Google's offer was rejected by Groupon, Google responded by creating its own program, called Google Offers, first tested in Portland (Oregon), San

Adlibrium
Amazon
AT&T
CrowdCut
DealFind
Dealicious
DealOn
DealSwarm
DoodleDeals
DoubleTake Deals
FabFind
Facebook Deals
Foursquare
GoDailyDeals
Groupon
Kijiji's Daily Deals
LivingSocial
LocalTwist
New York Times Limited
Playboy
RelishNYC
San Francisco Chronicle
Thrupon
WABC radio
Wagjag
Yelp
YourBestDeals

Fig. 3.1 Examples of companies offering e-coupons

Francisco, and New York in 2011, with expansion to 40 US cities by 2013 (Restaurant Hospitality 2011d). Groupon's largest competitor so far is the start-up firm, LivingSocial, which became a major player after receiving $400 million in venture capital from Amazon and other venture firms. LivingSocial is in the process of expanding into 400 cities. Meanwhile, AT&T is using the infrastructure of *The Yellow Pages*, which it owns, to build its own coupon website. Another competitor, Thrupon, uses the tagline "Groupon works for Groupon, Thrupon works for you" – referring to the fact that it retains a smaller portion of the coupon revenue than Groupon does. Google Offers also retains a lower portion of revenue (35 %), and this has caused Groupon to begin selectively reducing its cut of the revenue, now averaging 42 % instead of the original 50 %, in order to remain competitive in various submarkets. A list of some of the competitors is given in Fig. 3.1. Competition is fierce, and some restaurants report receiving two or three calls a day from the various e-coupon providers (Siegler 2010).

Groupon has experienced some other challenges in the marketplace. The novelty of e-coupons started to wear off in late 2011. Some restaurants and other businesses offering Groupons have reported that coupon customers treat them as a discount provider, returning at a later time to bargain shop for cut-rate deals. Some restaurateurs are concerned about the lack of demonstrated loyalty among coupon users; Groupon users, they believe, simply patronize the next restaurant that has a coupon to offer. To the contrary, however, Technomic's Online Daily Deal Report (2011) indicated that 67 % of consumers later returned to the restaurant without a coupon, 83 % recommended the restaurant to family or friends, 34 % posted a review on a reviewing website such as Yelp, and 25 % wrote about the restaurant on their Facebook page. Another commonly expressed complaint is that many Groupon customers are hard to please, e.g. writing negative reviews on Yelp afterwards, perhaps unfairly since the restaurant might not have been at its best while accommodating the large increase in business caused by the coupon promotion.

Groupon has faced some other external challenges as well. It has received sharp criticism in Europe that its advertisements are misleading (Yiannopoulos 2011). In 2011 it was forced to suspend its offers of coupons involving alcoholic beverages in Massachusetts restaurants because state law prohibits discounts on alcoholic beverages (Boston Globe 2011). Most Groupons expire after 2 or 3 months; Groupon was sued in federal court under the Credit Card Accountability Responsibility and Disclosure Act, which requires gift certificates to be valid for no less than 5 years. Groupon agreed to return the amount that the customer had paid for the coupon (not its redemption value) if the coupon was returned after it expired but before 5 years had passed (Lansu 2011). A survey of 400 businesses by Susquehanna Financial Group and Yipit released in early 2012 indicates that 52 % of businesses that had run a campaign with one of the e-coupon companies were not planning on another daily deal within the next 6 months, while another 24 % were planning only one deal in that time. Thus repeat business was expected to be hard to come by for Groupon, requiring the constant acquisition of new business in order to grow the company (Restaurant Hospitality 2011c; Sloane 2012). Several companies have gone into business to resell Groupon coupons for people who purchase them but find they cannot use them in time. The leader in this reselling business was Lifesta, but it has already gone out of business. DealsGoRound, CoupRecoup, and CityPockets are still in the reselling business as of this writing. Groupon discourages but does not prohibit reselling websites, expressing concern about the authenticity of the coupons. Generally, resale leads to a higher percentage of coupon redemption, which hurts the bottom line of the restaurants, which otherwise expect some percentage (typically about 20 %) of the coupons never to be redeemed.

There have also been challenges inside the company. The public relations chief and two chief operating officers resigned during 2011. Mason got into trouble with financial regulators. He wrote a long email message to all employees, responding to criticisms made of the company. When this email was leaked to the press, it was taken as a public statement, which is not permitted during the silent period preceding a public offering. The company has also been questioned about its accounting practices: for reporting as revenue the full amount of coupon receipts, even though half the funds are destined for the vendors; also for using the ACSOI (Adjusted Consolidated Segment Operating Income) accounting method, which leaves out marketing and acquisition expenses, thus making the company seem more profitable (Pepitone 2011). In response to these criticisms, the company returned to standard accounting practices, which caused it to restate earnings at less than half what it first reported. This new accounting showed that the company sustained an operating loss of $420 million in 2010. Other critics have noted the large amounts of capital removed from the operating budget to pay out dividends to Lefkosky and other early investors as well as the high debt-to-capital ratio (e.g., triple the debt-to-capital ratio of another startup firm, LinkedIn).

Mason has not always been the most careful or sensitive of spokesmen for the company. He is reported to have said: "We [Groupon] don't do shooting ranges, abortion clinics – we wouldn't do strip clubs or liposuction, things like that" (quoted in Williams 2010). One online reply shows the anger over this remark:

> Mason's lack of understanding of abortion is clear – the fact that he thinks any clinic would WANT to put a coupon out there for their life-changing service is grotesque. I am pro-choice (Is Mr. Mason? His millions of female shoppers might like to know if he supports the right of choice) … Two, to equate shooting ranges with abortion clinics, strip clubs and liposuction – I don't understand. Where's the correlation? Shooting ranges are a family-friendly activity wrapped up in a constitutional right of the great US of A. (Williams 2010; also see Dunham 2010)

Many industry watchers believe that Groupon made a bad business decision in turning down the Google offer and that the company is in serious financial trouble. Share prices dropped significantly in late 2011, as investors began to worry about whether Groupon had a viable business plan. Around this time, various independent auditors and industry watchers made dire predictions (Catanach and Ketz 2011; Conlin 2011; Foley 2011). Grumpy Old Accountants pronounced that the company was technically insolvent, with negative shareholder equity, and predicted that the company has a high probability of failure. PrivCo (2011) stated "Good headline, business still rotten." Industry watcher Conor Sen (2011) said, "it's operating like a Ponzi scheme that needs constant infusions of cash to stay afloat as it's hemorrhaging money." Fortune writer Robert Hof (2011) asked if this was Groupon's Webvan moment – referring to the dot-com start-up that lost a billion dollars (described in Chap. 2).

Despite these criticisms, the Security and Exchange Commission permitted Groupon to proceed with its initial public offering (Greenwood 2012; Parnell 2011; Raice 2011). Even as more than 170 of its direct competitors have gone out of business, Groupon continues to grow, with 142 million subscribers around the world (as of late 2011) and sales of over half a billion dollars in the fourth quarter of that year. Nevertheless, in 2012 the sales price of its shares continued to drop, from $20 at the time of the initial offering to under $4 by late 2012 (Wong 2012). Investors continue to worry about the long-term viability of the online coupon business model and Groupon's ability to sustain revenue growth. The Security and Exchange Commission spent most of 2012 scrutinizing the company's accounting and reporting practices, being especially concerned about the company's cash flow, given the full refund policy for its new businesses such as Groupon Getaways. Thus, as 2012 ended, the long-term viability of Groupon was very much in question. In February 2013 Mason was ousted by the board of directors, with the company having lost more than three-quarters of its value since 2011.

3.3   Case 3: Yelp

There are numerous websites where one can review restaurants online, including for example Yelp and Foursquare. Prior to online reviews, people generally found reviews in printed guidebooks. This genre became popular in the nineteenth century, when transportation technologies and increased private wealth made it easier for people to travel internationally. There was a major impetus to the publication of guide books between the two world wars on account of the increased availability of automobiles to middle class Americans. The most popular guidebook in America between the two world wars was *Adventures in Good Eating* (1936), written by the travelling salesman Duncan Hines. After the Second World War, many new guidebooks appeared in America, including popular series published by Eugene Fodor and Arthur Frommer. In recent years, the four most important guidebooks in American have been those from the Mobil oil company (now published by Forbes), the American Automobile Association, Michelin, and Zagat (purchased by Google in 2011).

It is useful to discuss the Michelin and Zagat guidebooks briefly, for they offer a contrast to Yelp's reviewing model, which we discuss in detail below. The Michelin guides were first published in France in 1900 but only introduced into the United States in 2005. Michelin reviews only the very highest quality restaurants. Its reviews are written by professional reviewers employed the company, who visit the restaurants anonymously to gather information for their reviews. The Zagat guides were created in 1979 by lawyers Tim and Nina Zagat, a husband and wife who quit their day jobs to build their guidebook company. Zagat collects reviews of restaurants from a wide set of volunteers. These reviews are severely edited and rendered into numerical scores on food, cost, service, and ambience – and presented together with short quotations from a few of the individual reviews. While Zagat reviews a larger number of restaurants than Michelin does, still it is only a minority of restaurants that receive reviews. Like Michelin, Zagat's focus is on higher end restaurants (Breidbart 1999; Leonhardt 2003; Willett 2012).

Yelp provides an interesting contrast to Zagat in the way that it carries out its reviewing. Yelp was created in San Francisco in 2004 by Jeremy Stoppelman and Russel Simmons, two friends who had worked together as software engineers at PayPal, and who cashed out when PayPal was sold to eBay (Foremski 2004). These friends began to look for something else to do. The creation myth is that Stoppelman was seeking a doctor but did not know how to find a good one. He conjured a convoluted scheme of emailing friends to ask for recommendations and then placing their responses on a public website where all could read them. Max Levchin, one of the cofounders of PayPal, provided $1 million to develop this email recommendation system. The system did not work out, but in the process of trying it out, Stoppelman and Simmons found that people were writing reviews for fun, and thus the basic idea for Yelp was born. The goal of Yelp is to allow crowds of people – almost anyone who wants to contribute – to review any kind of business that has a street address. While there are reviews of all kinds of businesses – bookstores and hair cutting salons, for example – the largest category of reviews are of restaurants. As Stoppelman explained: "Yelp just democratizes the reputation of a business... Rather than a single arbiter of taste, it's hundreds of people saying whether they like the business or not" (as quoted in Sutel 2007).

An entry for a typical business includes its average rating on Yelp's five-point rating scale, basic information about the business (name, category, business hours, telephone number, accessibility, and parking situation), and written reviews of the business (Graham 2007; Hansell 2008; McNeil 2008; Restaurants and Institutions 2007). Most reviews are shown on Yelp's website, but there is an automatically applied algorithm that filters out reviews that are deemed suspicious. Examples of suspicious reviews are ones that are overly supportive as though they may have been written by the friends and family of the business owner, or overly negative ones that seem as though they may have been written by competitors or people who hold a grudge against the business owner such as disgruntled former employees. Reviews that are otherwise not helpful are also filtered out. The filtered reviews are not expunged but instead appear on a page that is hidden unless the reader seeks it out.

Profiles are given of the reviewers, which can help readers discriminate among the possibly hundreds of reviews of a single restaurant. Readers can mark individual reviews as useful, funny, or cool; and a tally is kept of how many of each of these particular kudos each individual reviewer has received. Individual reviewers are also given credit for being the first to review a particular business, or for having written the best review of the day. Individual reviewers can also have a friends group, just as on Facebook or other social networking websites, where their reviews are readily available to their friends and they can chat with them (Porter 2008). Individual reviewers can also have fans, people who remain anonymous to the reviewer but who follow this individual's reviews. The profile page also contains photos, bookmarks, other content, and information about how long the individual has been a member of Yelp. The intention is to provide enough information that a reader can get a sense of the reviewer's interests and background, but not enough information that the reviewer can be personally identified. Yelp does not want a business owner who has received a negative review to be able to personally identify the reviewer. This privacy goal is difficult to achieve.

One problem for the reader is that having hundreds of reviews of a particular restaurant can offer an overload of information. Yelp has addressed this issue by building up a small core of reviewers who post reviews frequently, write particularly engaging reviews, or have large groups of friends and fans. This core is known as the Yelp Elite Squad (McCarthy 2007; Owyang 2008; Stross 2008). Being selected for the Elite Squad is generally regarded by the regular users of Yelp as a considerable honor, and the reviews written by members of the Elite Squad tend to be read much more often than those written by others. Yelp pampers its Elite Squad members, giving them special recognition online, offering them their own newsletter, and holding in-person social events exclusively for them (free dinners, open bars, etc.) in order to build community among these select reviewers. The *New York Times* profiled one Yelp reviewer (Megan Cress, known online as Megan C.) who, as of 2008, had written more than 300 reviews (95 of them "firsts"), and who had 957 friends and 151 fans on the Yelp website (McNeil 2008). [A profile of a different Yelp Elite member can be found in Chap. 1.]

The social networking aspects of the website and the success of the Yelp Elite Squad have contributed to Yelp's popularity. One Yelp Elite member (Ellen M. from Chicago) wrote: "I love to write reviews, but I think the social networking and interaction with yelp friends is what really compels me to continue." Yelp has much higher online traffic than other online restaurant reviewing websites. As of 2011, it had published 18 million reviews of businesses and was receiving 50 million unique visitors to the website each month.

Yelp's founders, Stoppelman and Simmons, were interested in social networking, and they used it as the foundation for their business:

> Actually, Stoppelman and Simmons weren't just looking for a new doctor. A pair of unrepentant party boys … they were in a perpetual search for the greatest restaurants and clubs in San Francisco. To get Yelp off the ground, they decided to mix business and pleasure, and started hosting Yelp parties at local establishments … (Flickr is littered with raucous snapshots from Yelp events featuring bar dancing and endless trains of women hanging all over the co-founders.) More important, the revelry got people writing reviews, building up the website's content.
>
> Today Yelpers seem to live on the website, messaging one another about their social lives, reacting to reviews, and planning get-togethers. That's the social-networking part. As is the case on most social networks, Yelp is rife with self-conscious patter. But there's a point to all the yammering: finding cool stuff that's not too far away. It's a mission everyone seems to take seriously. (O'Brien 2007)

It is instructive to compare Yelp and Zagat (as Zagat operated prior to its acquisition by Google in 2011). In Yelp the ratings and comments of every reviewer appear in complete form and are attributed to the particular reviewer (unless they have been filtered), while Zagat only publishes snippets from a few of the reviewers of a particular restaurant – and those reviewers are not identified. In Yelp, readers have free access to all reviews, whereas most of Zagat's ratings and reviews were available only to subscribers who paid for online service. Yelp accepts reviews about any restaurant and posts them, while Zagat accepts comments about any restaurant but only gives a rating and posts an entry on a small subset of the restaurants, as selected by its editorial staff. Yelp's web traffic is large and continues to grow rapidly, while Zagat's web traffic is much smaller and continues to grow at a moderate pace.

While this chapter emphasizes the aspects of Yelp as a restaurant reviewing website, if one focuses more on the business model, one can see Yelp's business as being one of capturing the local advertising market, which one scholar has valued at $100 billion annually (O'Brien 2007; also see Sutel 2007). O'Brien notes that Yelp's model, of using crowdsourcing to produce business reviews, is only one of several possible business models. Others include a directory model (e.g. used by most traditional businesses), in which a large sales force attempts to persuade business owners to buy additional and more expensive ads; a proprietary model (e.g. CitySearch), in which a company creates proprietary content and sells ads in support of it; or a search engine model (e.g. Google), in which a company crawls the web and tries to automate the sales of advertisements to those businesses uncovered in the crawl. When viewed in this way, Yelp can be seen in trying to take over some of the market currently served by *The Yellow Pages*. This analysis squares with another part of Yelp's creation myth, in which Stoppelman had become fascinated with the way in which Craigslist had undermined traditional classified advertising in newspapers.

Over time, the company needed extra funding primarily for office space and additional sales staff as it expanded into new cities and countries. As of 2010, the company employed 600 workers and was sponsoring 900 events per year for its Elite Squad (Stoppelman 2010). The company has successfully raised funds in several rounds of funding. In addition to the initial $1 million from Max Levchin, the company received $5 million in 2005 from Bessemer Ventures, a sponsor of Skype. Another round of funding came in 2006 from Benchmark Capital, a sponsor of eBay. A third round of funding in the amount of $15 million came from DAG Ventures in 2008, a sponsor of Friendster (McCarthy 2008; McCarthy 2008).

In 2009 Google came close to purchasing Yelp for more than $500 million (Arrington 2009). When the deal fell through, Google began to develop its own Google Places business in competition with Yelp. Google was strongly criticized for using Yelp reviews on its Google Places searches without a licensing agreement with Yelp, and this was partly rectified when Google purchased Zagat, which gave it a legitimate source of reviewer content to post (Barth 2011; Chan and McBride 2011; de la Merced et al. 2011; Kincaid 2011). In late 2011 Yelp filed with the Securities and Exchange Commission to raise up to $100 million in capital. The company had a successful public offering, with the stock value rising 64 % on its first day of trading in March 2012 (Evangelista 2012; Tan and Kucera 2012). However, the company has not seen a profit since 2007, and despite a large increase in revenue it lost more than $16 million in 2011.

One of the company's successful strategies was to focus its early efforts on building critical mass in the San Francisco area before moving into other markets. There are over 19,000 reviews of restaurants in the San Francisco area, and there is a general sense that Yelp has rather complete and thorough coverage of the San Francisco restaurant scene (Lannin 2011; Maddan 2006). As Yelp first expanded into other cities, such as Los Angeles, San Diego, Chicago, Boston, New York, and Washington, it adopted a practice employed by several of its competitors (including Epinions, Judy's Book, and InsiderPages) to pay people to write reviews so that there was sufficient content on businesses in the expansion cities to attract users (San Francisco Business Times 2006). This method did not work well for Yelp, so it dropped this practice in 2008 to focus on building up Elite Squad presence in these expansion locales, using in-person events to build the community of reviewers. The company waited until 2008 to expand outside the United States – first to Canada and then to Western Europe and Australia (Mawby 2011; Schonfeld 2011). The low-profile advertising style was intended not to offend reviewers or readers as being too commercial. Yelp posts no banner ads, for example. Most of the advertising revenue comes instead from business listings and in particular from providing preferred placement to companies on the search pages. To keep the focus on quality businesses, Yelp only accepts ads from businesses that received an average score of at least three stars (on its five-star scale). One can access Yelp by smartphone as well as computer (Jensen 2008).

Yelp has had a number of competitors. Los Angeles-based CitySearch, which generally publishes more moderate reviews than Yelp, was the market leader when Yelp entered the field, but Yelp surpassed CitySearch within 2 years. Yelp learned some of its social network business model from Epinions, which differs from Yelp in that its website allows comments not only on restaurants or other businesses, but also on specific products (e.g., vegetable peelers) and on non-profit institutions (e.g., colleges). The interaction between reviewers and other consumers is a valued feature of Epinions: "The cool part of Epinions ... is waiting and seeing how people rate your review (quoted in Frauenfelder 2000). Epinion reviewers could sign up for email to alert them when somebody has rated their reviews. Some of the early entrants went out of business quickly, such as Seattle-based Judy's Book and Intuit's Zipingo (Arrington 2007). Other competitors included general-purpose InsiderPages, travel websites such as TripAdvisor, Gogobot, and Afar, and restaurant website MenuPages (Yu 2011). The blending of Yelp's online reviewing with its in-person social events is similar to the online social event calendar Going, the online website for finding dates and drinking buddies I'm In Like With You, and the social action mobilization website Meetup. Eat-A-Rama, which runs as an application of Facebook for restaurant discovery, might also be considered a competitor (Restaurant Hospitality 2008, 2011a).

Restaurants, and businesses in general, have a love-hate relationship with Yelp (Trapunski 2011). Favorable reviews on Yelp can generate extensive customer traffic to the restaurant. But unfavorable reviews can be harmful to custom, and the restaurant owner might not even be aware of the reviews – or even of Yelp itself. In 2008 Yelp wanted to improve its relations with businesses, so it provided them with some new tools: a system to send messages to reviewers, a way to find out how many people viewed their business page on Yelp, a tool to update their business entry more easily and quickly, and a way to receive email notification when new reviews of their business were posted (Duxbury 2008; Gonzalez 2008).

There are speculations in the literature that some businesses have tried to use Yelp to bolster their business by surreptitiously publishing their own favorable reviews or paying a service such as Fiverr to write and post favorable reviews on behalf of the business (Streitfeld 2011a). Yelp wants to prevent this from happening, but the large number of postings means that Yelp's first line of defense has to be an automated filtering tool, and at times spurious reviews pass muster under this automated scrutiny (Leung 2010; Lewis 2011; Lubin 2012). Writing such reviews is against Federal Trade Commission rules passed in 2009, which ban undisclosed paid endorsements, but a small business owner is unlikely to get caught. There are reports of business owners offering discounts or free products to customers who write good reviews on their behalf. It is unclear how widespread these activities are.

What is much more common are complaints by business owners about pressure tactics from the Yelp sales force to buy advertising to help them manage negative reviews. In one well-publicized court case, Dr. Gregory Perrault, the owner of the Cats and Dogs Animal Hospital in Long Beach, California, sued Yelp for extortion (Zetter 2010). The story begins with a request to Yelp from Perrault to pull a negative review. Yelp obliged because the incident that was discussed in the review had occurred more than 18 months prior to the posting of the review and thus did not meet Yelp's timeliness policy. However, soon a new negative review appeared on Yelp by someone signed on as Kay K., and shortly after this, Yelp sales staff called on Perrault to sell him advertising, with the assurance that they could help him to manage negative reviews like the one from Kay K. Perrault declined the advertising contract, and soon the 18+ month old review reappeared on Yelp, as did another negative review by Kay K. that made personal attacks on Perrault. Perrault complained, but Yelp left the offending reviews online. At this point Perrault sued Yelp for extortion and fraudulent business practices, claiming that Yelp sales people used high-pressure sales tactics on him, offering to remove or place negative ads in a less accessible place on the Yelp website if he would purchase a 1-year advertising contract at the cost of $300 per month.

The case was amended in 2010 to a class action suit when nine small business owners, including a Chicago bakery, a Washington, DC restaurant, and a California furniture store joined Perrault as plaintiffs (Ali 2010). US District Court Judge Marilyn Patel dismissed the extortion charge against Yelp in 2011 because the communications from Yelp brought into evidence did not meet the criteria for being extortion (McCarty 2011). Yelp CEO Jeremy Stoppelman used combative language in rebutting these charges. For example, on the official Yelp blog he wrote: "Last year [2010], a few small businesses from among the 20 million or so in the United States filed misguided lawsuits against Yelp...."

Although the lawsuit was settled in Yelp's favor, the suit brought to light a number of other complaints from journalists and business owners about Yelp's sales practices (Falconer 2011; Metz 2008). CBS Channel 5 in San Francisco, the British online science-technology journal *The Register*, *Wired* online, and the newspaper *East Bay Express* all ran articles reporting on complaints about Yelp similar to those lodged by Perrault (Pavini 2008). In an *East Bay Express* article Kathleen Richards reported that she had heard from six small business owners who claimed that Yelp sales staff had promised to move or remove negative reviews if the business would sign up for an advertising contract, and another six reported that positive reviews disappeared or negative ones appeared after they refused a Yelp advertising contract (Richards 2009a). Stoppelman derided the article, criticizing it for using anonymous sources. Richards then wrote a follow-up article in the same newspaper that identified two sources, the owner of M&M Auto Werkes in Campbell, California and Calvin Gee, the owner of Haight Street Dental in San Francisco, who both complained about similar treatment (Richards 2009b).

Several of the online articles had comments posted from business owners complaining about Yelp's sales tactics. Here is a comment to the *Wired* article, posted by Gloria Kardong, an M.D. in Palo Alto, California:

> I am a business owner. I had all 5 star unsolicited reviews from my clients. Then I started to receive aggressive advertising calls. I refused to advertise in exchange for good ratings because I don't need to do so and because it is unethical. I also told them that everyone knows about the extortion yelp perpetrates on small business owners who don't advertise with them. One week later, I received a one star rating. My 10 and 20 year clients who gave me all 5 star reviews were put into the filtered category and the one star rating was left behind. When I complained about this and made the connection between my refusal to advertise with them and the plummeting rating from 5 star to 1 star led to them filtering another 5 star rating. I also asked them to remove the 1 star rating because it was a fabrication that violated their own policies. They responded by filtering another 5 star rating. I have ten 5 star ratings and one 1 star rating. They list only 2 of these 5 star ratings along with the one star rating, cutting my overall rating in half. All of you who are complaining below are inviting more retaliation. Thanks for speaking up anyway.

Yelp has also been sued for patent infringement by Earthcomber, a Chicago-based company that received a patent in 2006 for a process to use software based on global positioning system technology to match a person to a nearby place of interest (an event or a particular kind of business). Earthcomber appears to be a patent troll because it also sued OpenTable, Groupon, and ten online real estate businesses including Zillow and Trulia on the same grounds (Roberts 2012). When sued, Zagat licensed with Earthcomber, causing Earthcomer to become more aggressive in its lawsuits.

In another well-covered story, one can see how the tension between negative reviewers on Yelp and small business owners can erupt in violence. Sean C. wrote a negative review on Yelp of Ocean Avenue Bookstore in San Francisco. The book-store owner, Diane Goodman, wrote a nasty message to Sean C. using the tool that Yelp provides enabling a business owner to write a one-time email reply to a reviewer. Sean C. posted screen shots of Goodman's message on flickr. Goodman was able to ascertain Sean C.'s identify from his flickr user profile and came to his house with a gun. They had a tussle when she tried to force herself in his house, and during the altercation she fell down several stairs. Police charged her with battery, and Sean C. took out a restraining order. The bookstore subsequently closed (NBC Bay Area 2010; Yoo 2009).

Some restaurant owners are deeply unhappy with Yelp and the nature of the reviewing process, viewing Yelpers as "a fork-waving mob of know-nothings" (McNeil 2008). Consider the case of well-known Chicago chef Graham Elliot Bowles (known professionally by his first two names), who has appeared on several television cooking shows and holds a Michelin star for one of his restaurants (Stein 2010). Bowles has been kicked off of Yelp three times for replying to Yelp reviews that he believed to be factually incorrect. His disgust with Yelp was cemented when he received a Yelp review of his new restaurant (Grahamwich) before it opened. Somebody had read in a Chicago magazine that Bowles was opening the new restaurant, walked down to the location to try it out but found it had not opened yet.

So the person gave a negative rating (1 star out of 5) and referred it his review of "a pleasant walk ruined" (Brion 2010).

Four other general criticisms have been lodged in the press about Yelp. The first is that Yelp's reviewing process can discourage food innovation or menus that are targeted at a narrowly focused audience. Apparently an extra star in the Yelp rating system translates to an average of 9 % of additional business. So a restaurateur may be risk averse in deciding what to serve, so as to appeal to the typical Yelp reviewer. Second, Yelp might harm restaurants by premature reviewing. Most restaurants take a while to get the kinks out of their operations when they first open. In earlier times, there was typically a grace period before the first review visits were made, but Yelp encourages people to be the first to review a business and restaurants often receive Yelp reviewers on their opening night. Third, Yelp may damage the thrill of exploring a new restaurant. A person might decide not to go to the restaurant after reading the reviews. Even if she does decide to go, she might know so much about the restaurant from the reviews that there is little new to experience when she gets there. This is an instance of what author Edward Tenner calls counter-serendipity, where advanced knowledge of something reduces opportunities for fortuitous experiences (Doig 2012). Finally, Yelp reviews may also represent a segment of the population that is mismatched with the interests of a particular reader or the target audience of the restaurant. The reviewing population is generally young (Yelp says that 42 % are between the ages of 18 and 34) and without children (63 %). The reviewers tend to be value-conscious; they are not generally willing to pay premium prices for premium products. Thus, reviewers of a restaurant designed to cater to an older, wealthier clientele might not review favorably on Yelp. Because of the nature of its reviewers, Yelp reviewing ignores the fact of market segmentation and applies its standards to segments that might have different values (Simester 2011).

3.4 Conclusions

There is no doubt that the Internet provides extraordinary opportunities both for enhancing traditional commerce and for creating powerful new business models for such tasks as making restaurant reservations, distributing coupons, and reviewing restaurants. Through examples of three major companies from the online food industry, we have identified some of the problems that can arise for customers or traditional businesses from their interactions with online companies. It is not clear that these dangers are inherent in online businesses. One can imagine a different pricing model for restaurant reservations, an online coupon company that protected its small business clients better, or less aggressive sales practices by restaurant reviewing companies. However, these three cases provide a cautionary tale about the dangers of the powers that the Internet affords. These dangers have been illustrated here in the context of the food industry, but they apply much more widely to other business areas.

References

Ali S (2010) Small businesses join lawsuit against Yelp. Digits (blog), Wall Street Journal. http://blogs.wsj.com/digits/2010/03/17/small-businesses-join-lawsuit-against-yelp. 17 March 2010, Accessed 7 June 2013

Arrington M (2007) Judy's book to shut down; Yelp is the last of the local review sites still standing. Tech Crunch. http://techcrunch.com/2007/10/23/judys-book-to-shut-down-yelp-is-the-last-of-the-local-review-sites-still-standing. 23 Oct 2007, Accessed 7 June 2013

Arrington M (2009) Google in discussions to acquire Yelp for a half billion dollars or more. Tech Crunch. http://techcrunch.com/2009/12/17/google-acquire-buy-yelp. 17 Dec 2009, Accessed 7 June 2013

Arrington M (2010) Getting to the bottom of the crazy Yahoo-Groupon rumors. Tech Crunch. http://techcrunch.com/2010/10/16/getting-to-the-bottom-of-the-crazy-yahoo-groupon-rumors. 16 Oct 2010, Accessed 7 June 2013

Aspray W, Cortada JW, Ocepek MG, Royer G (2013) Against internet exceptionalism: Understanding OpernTable through a traditional business history of technology approach. Journal of Internet Business 11

Austin S (2009) Venture firms make reservations to sell OpenTable stock. Venture Capital Dispatch (blog), Wall Street Journal. http://blogs.wsj.com/venturecapital/2009/09/22/venture-firms-make-reservations-to-sell-opentable-stock. 22 Sept 2009, Accessed 15 June 2013

Barth C (2011) Google paid $151 million to get Zagat, ditch Yelp. Intelligent Investing (blog), Forbes. http://www.forbes.com/sites/chrisbarth/2011/10/27/google-paid-151-million-to-get-zagat-ditch-yelp. 27 Oct 2011, Accessed 15 June 2013

BBC (2012) Groupon voucher deal 'cost Hove firm thousands.' BBC News Sussex. http://www.bbc.co.uk/news/uk-england-sussex-16679683. 23 Jan 2012, Accessed 15 June 2013

Bloomberg Businessweek (2012) Opentable, Inc. http://investing.businessweek.com/research/stocks/snapshot/snapshot.asp?ticker=OPEN:US. Accessed 9 Feb 2012

Boston Globe (2011) Groupon suspends discount drink policy in Mass. http://www.boston.com/business/ticker/2011/03/groupon_suspend.html. 18 March 2011, Accessed 15 June 2013

Bowman RD (1985) Profit on the dotted line: Coupons and rebates. 2nd edn. Commerce Communications, Chicago

Breidbart S (1999) Flaws in restaurant guides. Letter to the editor, New York Times. http://travel.nytimes.com/top/reference/timestopics/subjects/t/travel_and_vacations/index.html?query=BREIDBART,%20SHAUN&field=per&match=exact. 22 Dec 1999, Accessed 15 June 2013

Brion R (2010) Graham Elliot's unopened resto gets a negative Yelp review. Eater. http://eater.com/archives/2010/09/01/graham-elliots-grahamwich-gets-negative-yelp-review.php. 1 Sept 2010, Accessed 15 June 2013

Burkeman O (2011) There's nothing new in bulk buying and negotiating discounts. Guardian. 11 June 2011

Carrera I (2011) Aftermarket Groupon sites eat into revenues. New York Times. http://www.nytimes.com/2011/09/16/us/aftermarket-groupon-sites-eat-into-revenues.html. 15 Sept 2011, Accessed 15 June 2013

Catanach AH Jr, Ketz JE (2011) Groupon is technically insolvent. Grumpy Old Accountants (blog). http://blogs.smeal.psu.edu/grumpyoldaccountants/archives/362. 21 Oct 2011, Accessed 15 June 2013

Chan E, McBride S (2011) Google dines on Zagat to bolster local guides. Toronto National Post. 9 Sept 2011

Coburn MF (2010) On Groupon and its founder, Andrew Mason. ChicagoMag.com. http://www.chicagomag.com/Chicago-Magazine/August-2010/On-Groupon-and-its-founder-Andrew-Mason. Aug 2010, Accessed 15 June 2013

Cohen DL (2009) Virtual 'tipping point' leverages group deals. Reuters. http://www.reuters.com/article/2009/06/10/us-groupon-idUSTRE5592K720090610. 10 June 2009, Accessed 15 June 2013

Conlin M (2011) Groupon's fall to earth swifter than its fast rise. Associated Press. http://news.yahoo.com/groupons-fall-earth-swifter-fast-rise-184713324.html. 21 Oct 2011, Accessed 15 June 2013

Davis H (2012) Local restaurants embrace phone app allowing customers to make reservations online. WDBJ7.com. http://www.wdbj7.com/news/wdbj7-local-restaurants-embrace-app-that-allows-customers-to-book-reservations-online-20120228,0,6072332.story. 28 Feb 2012, Accessed 15 June 2013

de la Merced MJ, Lieber R, Miller CC (2011) In a twist, Google reviews Zagat, and decides to bite. Deal Book (blog), New York Times. http://dealbook.nytimes.com/2011/09/08/google-to-buy-zagat/?ref=timzagat. 8 Sept 2011, Accessed 15 June 2013

Dholakia UM (2011) How businesses fare with daily deals: A multi-site analysis of Groupon, Livingsocial, Opentable, Travelzoo, and BuyWithMe promotions. http://ssrn.com/abstract=1863466. 13 June 2011, Accessed 15 June 2013

Doig W (2012) How Yelp destroyed the thrill of exploring. Salon.com. http://www.salon.com/2012/01/28/how_yelp_destroyed_the_thrill_of_exploring. 28 Jan 2012, Accessed 15 June 2013

Dunham J (2010) Marketing to women: Groupon or groupoff? Lipstick Economy (blog). http://jamiedunham.wordpress.com/?s=groupoff. 8 Oct 2010, Accessed 15 June 2013

Duxbury S (2008) Restaurants learn to Yelp. San Francisco Business Times. http://www.bizjournals.com/sanfrancisco/stories/2008/06/30/story1.html?page=all. 29 June 2008, Accessed 15 June 2013

Evangelista B (2012) Yelp shares soar nearly 64% in IPO. San Francisco Chronicle. http://www.sfgate.com/news/article/Yelp-shares-soar-nearly-64-in-IPO-3378737.php. 3 March 2012, Accessed 15 June 2013

Falconer J (2011) Yelp extortion lawsuit dismissed but not dead. The Next Web. http://thenextweb.com/insider/2011/03/30/yelp-extortion-lawsuit-dismissed-but-not-dead. 30 March 2011, Accessed 15 June 2013

Foodie Buddha (2011) UrbanSpoon Rez hits Atlanta. http://www.foodiebuddha.com/2011/04/22/urbanspoon-rez-hits-atlanta-online-restaurant-reservations. 22 April 2011, Accessed 15 June 2013

Foley S (2011) Is it time to discount the king of coupons? Independent (London). http://www.independent.co.uk/news/business/analysis-and-features/is-it-time-to-discount-the-kings-of-coupons-2278488.html. 4 May 2011, Accessed 15 June 2013

Foremski T (2004) Tech watch: Yelp! – PayPal co-founder pops first venture out of incubator. Silicon Valley Watcher. http://www.siliconvalleywatcher.com/MT-3.34-en/mt-search.cgi?search=mrlv+incubator&IncludeBlogs=6&limit=100. 12 Oct 2004, Accessed 15 June 2013

Frauenfelder M (2000) Revenge of the know-it-alls. Wired. http://www.wired.com/wired/archive/8.07/egoboo.html. July 2000, Accessed 15 June 2013

Fund Manager (2011) Online restaurant reservations just got a lot cheaper. Seeking Alpha. http://seekingalpha.com/article/307933-online-restaurant-reservations-just-got-a-lot-cheaper. 15 Nov 2011, Accessed 15 June 2013

Geron T (2009) IPO-ready OpenTable hit with suspiciously timed lawsuit. Venture Capital Dispatch (blog), Wall Street Journal. http://blogs.wsj.com/venturecapital/2009/05/19/ipo-ready-opentable-hit-with-suspiciously-timed-lawsuit. 19 May 2009, Accessed 15 June 2013

Gonzalez N (2008) Yelp lets businesses fight back. Tech Crunch. http://techcrunch.com/2008/04/28/yelp-lets-businesses-fight-back. 28 April 2008, Accessed 15 June 2013

Graham J (2007) 'Yelpers' review local businesses. USA Today. http://www.usatoday.com/tech/webguide/2007-06-12-yelp_N.htm. 12 June 2007, Accessed 21 Feb 2012

Greenwood A (2012) Groupon cancels daily deals, Attempts to close books on SEC investigation. National Monitor. http://pandithnews.com/2012/11/03/groupon-cancels-daily-deals-attempts-to-close-books-on-sec-investigation/. 3 Nov 2012, Accessed 15 June 2013

Hafner K (2007) Restaurant reservations go online. New York Times. http://www.nytimes.com/2007/06/18/business/18opentable.html?pagewanted=all. 18 June 2007, Accessed 15 June 2013

Halm M (2000) Online restaurant reservations really do work. CNN Tech, http://articles.cnn.com/2000-01-27/tech/reserve.food.online.idg_1_online-reservation-reservation-services-restaurant?_s=PM:TECH. 27 Jan 2000, Accessed 15 June 2013

Hansell S (2008) Why Yelp works. Bits (blog), New York Times. http://bits.blogs.nytimes.com/2008/05/12/why-yelp-works. 12 May 2008, Accessed 15 June 2013

Hof R (2011) Is this Groupon's Webvan moment? Forbes. http://www.forbes.com/sites/roberthof/2011/11/23/is-this-groupons-webvan-moment. 23 Nov 2011, Accessed 15 June 2013

Jensen MD (2008) Yelp bringing local reviews to your iPhone and the world. City Marketer. http://web.archive.org/web/20080730023444/http://citymarketer.com/category/yelp 5 June 2008, Accessed 22 Feb 2012

Joseph S (2006) Table for 2 is a click away – part II. Orlando Sentinel. http://press.opentable.com/releasedetail.cfm?ReleaseID=486908. 8 Sept 2006, Accessed 15 June 2013

Kimes SE (2009) How restaurant customers view online reservations (executive summary). Cornell Hospitality Reports 9(5). http://www.hotelschool.cornell.edu/research/chr/pubs/reports/abstract-15006.html. Accessed 5 March 2012

Kincaid J (2011) Google acquires Zagat to flesh out local reviews. Tech Crunch. http://techcrunch.com/2011/09/08/google-acquires-zagat-to-flesh-out-local-ratings. 8 Sept 2011, Accessed 15 June 2013

Lang JM (2006) Is the web really a sales builder? Restaurant Business. May 2006, 11–12

Lannin S (2011) Nearly 25% of bay area restaurant reservations made online; OpenTable tackles foreign markets. In the Peninsula (blog), San Francisco Chronicle. http://blog.sfgate.com/inthepeninsula/2011/04/21/nearly-25-of-bay-area-restaurant-reservations-made-online-opentable-tackles-foreign-markets. 21 April 2011, Accessed 15 June 2013

Lansu M (2011) Groupon sued over deal expiration dates. Chicago Sun-Times. http://www.sun-times.com/business/4083887-420/groupon-sued-over-deal-expiration-dates.html. 2 March 2011, Accessed 15 June 2013

Leonhardt D (2003) Everyone's a critic. New York Times. http://www.nytimes.com/2003/11/23/books/everyone-s-a-critic.html?pagewanted=all&src=pm. 23 Nov 2003, Accessed 15 June 2013

Leson N (2005) Risks and rewards of booking your table online. Seattle Times. http://community.seattletimes.nwsource.com/archive/?date=20050817&slug=taste17. 17 Aug 2005, Accessed 15 June 2013

Leung W (2010) Sniffing out fake restaurant reviews. Toronto Globe and Mail. 13 Oct 2010

Leung W (2011) Online coupon sites turn up heat on restaurants. Toronto Globe and Mail. 11 May 2011

Lewis R (2011) Beware rigged online reviews. CBS News. http://www.cbsnews.com/8301-502303_162-57318375/beware-rigged-online-reviews. 4 Nov 2011, Accessed 15 June 2013

Lieberman D (1999) Info superhighway has deals on wheels. USA Today. 15 Nov 1999

Livebookings (2011) Livebookings launches freebookings, a free online restaurant reservation system. Press release. http://www.livebookings.us/News/Livebookings_Launches_Freebookings_a_Free_Online_Reservation_Service_for_Restaurants. 10 Nov 2011, Accessed 15 June 2013

Localized USA (2012) OpenTable Inc (OPEN) shares upgraded to a 'overweight' rating by Barclays Capital (BCS) analysts. http://localizedusa.com/2012/01/09/opentable-inc-open-shares-upgraded-to-a-overweight-rating-by-barclays-capital-bcs-analysts. 9 Jan 2012, Accessed 15 June 2013

Lubin G (2012) The illegal way to improve your rating on Yelp. Business Insider. http://www.businessinsider.com/the-illegal-way-to-improve-your-rating-on-yelp-2012-2?utm_ source=feedburner&utm_medium=feed&utm_campaign=Feed%3A+typepad%2Falleyinsider %2Fsilicon_alley_insider+%28Silicon+Alley+Insider%29. 19 Feb 2012, Accessed 15 June 2013

MacMillan D, Galante J (2010) Groupon prankster Mason not joking in spurning Google. Bloomberg.com. http://www.bloomberg.com/news/2010-12-06/groupon-prankster-mason-not-joking-in-spurning-google-s-6-billion-offer.html. 6 Dec 2010, Accessed 15 June 2013

McCarthy C (2007) Social networks geared for offline success? CNET. http://news.cnet.com/ Social-networks-geared-for-offline-success/2100-1038_3-6192780.html. 22 June 2007, Accessed 15 June 2013

McCarthy C (2008) Yelp yanks another $15 million. CNET. http://news.cnet.com/8301-13577_3-9880723-36.html. 27 Feb 2008, Accessed 15 June 2013

McCarthy M (2008) A mighty Yelp! Review site gets $15M. Wired. http://www.wired.com/ epicenter/2008/02/a-mighty-yelp-r. 27 Feb 2008, Accessed 15 June 2013

McCarty B (2011) Class action lawsuit against yelp dropped, CEO calls filers 'small' and 'misguided.' The Next Web. http://thenextweb.com/insider/2011/10/27/class-action-lawsuit-against-yelp-dropped-ceo-calls-filers-small-and-misguided. 27 Oct 2011, Accessed 15 June 2013

McLaughlin K (2010) More ways to snag that restaurant table. Wall Street Journal. http://online. wsj.com/article/SB10001424052748703691804575254680739522068.html. 20 May 2010, Accessed 15 June 2013

McNeil DG Jr (2008) Eat and Tell. New York Times. http://www.nytimes.com/2008/11/05/ dining/05yelp.html?pagewanted=print. 4 Nov 2008, Accessed 15 June 2013

Maddan H (2006) Casting the net: Yelp is on the way. San Francisco Chronicle. http://www.sfgate. com/cgi-bin/article.cgi?f=/c/a/2006/06/18/LVGO9JDMDV1.DTL&ao=all. 18 June 2006, Accessed 15 June 2013

Mawby N (2011) Yelp is at hand when you need to know. Melbourne Herald Sun. http://www. heraldsun.com.au/business/yelp-is-at-hand-when-you-need-to-know/story-fn7j19iv-1226213592380. 5 Dec 2011, Accessed 15 June 2013

Metz C (2008) Yelp 'pay to play' pitch makes shops scream for help. The Register. http://www. theregister.co.uk/2008/08/13/yelp_sales_pitch. 13 Aug 2008, Accessed 15 June 2013

Morell K (2011) OpenTable founder Chuck Templeton on starting up. American Express Open Forum. http://www.openforum.com/articles/opentable-founder-chuck-templeton-on-starting-up. 26 Oct 2011, Accessed 15 June 2013

Mui YQ (2010) Group coupons can be too popular. Washington Post. http://www.washingtonpost. com/wp-dyn/content/article/2010/09/17/AR2010091706759.html. 18 Sept 2010, Accessed 15 June 2013

NBC Bay Area (2010) San Francisco bookstore accused of violent yelp confrontation. http://www. nbcbayarea.com/the-scene/shopping/San-Francisco-Bookstore-Has-an-Uneahlty-Yelp-Confrontation-68914367.html. 29 Jan 2010, Accessed 15 June 2013

O'Brien JM (2007) Business paradigm shifts and free tequila shots. CNN Money. http://money. cnn.com/magazines/fortune/fortune_archive/2007/07/23/100134489. 10 July 2007, Accessed 15 June 2013

O'Dell J (2010) Groupon goes international, buys Japanese and Russian clones. Mashable. http:// mashable.com/2010/08/17/groupon-international. 17 Aug 2010, Accessed 15 June 2013

Owyang J (2008) Understanding community leadership: An interview with a member of Yelp's 'elite.' Web Strategist (blog). http://www.web-strategist.com/blog/2008/05/27/understanding-community-leadership-an-interview-with-a-member-of-yelps-elite. 27 May 2008, Accessed 15 June 2013

Parnell BA (2011) Groupon snatches a great deal . . . on itself. The Register. http://www.theregister.co.uk/2011/11/04/groupon_ipo_price. 4 Nov 2011, Accessed 15 June 2013

Pavini J (2008) Business owners 'Yelp' about Internet ratings site. CBS 5 San Francisco. http:// web.archive.org/web/20080822011929/http://cbs5.com/wrapper_consumer/seenon/Yelp. Internet.ratings.2.787400.html. 5 Aug 2008, Accessed 15 June 2013

Pepitone J (2011) Groupon updates IPO filing, admits it's unprofitable. CNN Money. http://money. cnn.com/2011/08/10/technology/groupon_accounting. 10 Aug 2011, Accessed 15 June 2013

Perez S (2011) Adlibrium launches daily deals service where offers appear as mobile ads. Tech Crunch. http://techcrunch.com/2011/10/18/adlibrium-launches-daily-deals-service-where-offers-appear-as-mobile-ads. 18 Oct 2011, Accessed 15 June 2013

Porter J (2008) Social design patterns for reputation systems: An interview with yahoo's bryce gass (part II). Bokardo (blog). http://bokardo.com/archives/social-design-patterns-for-reputation-systems-two. 25 June 2008, Accessed 15 June 2013

Primack D (2011) Will Google try to fold OpenTable? CNN Money. http://finance.fortune.cnn.com/2011/09/08/will-google-try-to-fold-opentable. 8 Sept 2011, Accessed 15 June 2013

PrivCo (2011) Update for Groupon, Inc. http://www.privco.com/press/october-21-2011-groupons-3rd-quarter-financials-filed-this-morning-privcocom-analysis-all-key-business-metrics-took-quarterly-sequential-tumble-again. 21 Oct 2011, Accessed 15 June 2013

Raice S (2011) More trouble for Groupon IPO. Wall Street Journal. http://online.wsj.com/article/SB10001424053111903791504576589211214409214.html. 24 Sept 2011, Accessed 15 June 2013

Restaurant Hospitality (2008) Wisdom of the crowd. http://restaurant-hospitality.com/observer/wisdom_crowd. 1 March 2008, Accessed 15 June 2013

Restaurant Hospitality (2010) So, everybody loves OpenTable. Do you? http://restaurant-hospitality.com/trends/everybody-loves-opentable-table-1210. 1 Dec 2010, Accessed 15 June 2013

Restaurant Hospitality (2011a) Social media: Ignore it at your peril. http://restaurant-hospitality.com/news/social-media-ignor-peril-0207. 7 Feb 2011, Accessed 15 June 2013

Restaurant Hospitality (2011b) Groupon deals often succeed. http://restaurant-hospitality.com/news/groupon-deals-succeed-0611. 10 June 2011, Accessed 15 June 2013

Restaurant Hospitality (2011c) Has deal fatigue set in? http://restaurant-hospitality.com/trends/deal-fatigue-set-in-0611. 15 June 2011, Accessed 15 June 2013

Restaurant Hospitality (2011d) Google offers or Groupon: You make the call. http://restaurant-hospitality.com/trends/google-offers-or-groupon-0711. 18 July 2011, Accessed 15 June 2013

Restaurants and Institutions (2007) Yelping hands. 1 Jan 2007

Richards K (2009a) Yelp and the business of extortion 2.0. East Bay Express. http://www.eastbay-express.com/ebx/yelp-and-the-business-of-extortion-20/Content?oid=1176635. 18 Feb 2009, Accessed 15 June 2013

Richards K (2009b) Yelp extortion allegations stack up, East Bay Express. http://www.eastbayex-press.com/ebx/yelp-extortion-allegations-stack-up/Content?oid=1176984. 18 March 2009, Accessed 15 June 2013

Roberts JJ (2012) Updated: App maker uses GPS patent to sue Zillow, Trulia. PaidContent. http://paidcontent.org/article/419-app-maker-uses-gps-patent-to-sue-zillow-trulia. 20 Jan 2012, Accessed 15 June 2013

Rowe M (2009) What the web can do for you. Restaurant Hospitality. http://restaurant-hospitality.com/features/web-can-you-0609. 1 June 2009, Accessed 15 June 2013

San Francisco Business Times (2006) City guide Yelp raises $10M in second round. http://www.bizjournals.com/sanfrancisco/stories/2006/10/02/daily35.html. 5 Oct 2006, Accessed 15 June 2013

Schonfeld E (2011) Yelp moves to Spain as international traffic doubles. Tech Crunch. http://techcrunch.com/2011/05/11/yelp-spain. 11 May 2011, Accessed 15 June 2013

Sen C (2011) Groupon is effectively insolvent. Minyanville. http://www.minyanville.com/businessmarkets/articles/groupon-groupon-ipo-tech-stocks-linked/6/3/2011/id/34936. 3 June 2011, Accessed 15 June 2013

Sennett F (2012) Groupon's biggest deal ever. St. Martin's, New York

Siegler MG (2010) The other location shoe drops: Facebook Deals. Will it discount rivals? Tech Crunch. http://techcrunch.com/2010/11/03/facebook-deals. 3 Nov 2010, Accessed 15 June 2013

Silicon Valley/San Jose Business Journal (2008) OpenTable's free iPhone app finds nearby dining reservations. http://www.bizjournals.com/sanjose/stories/2008/11/17/daily2.html. 17 Nov 2008, Accessed 15 June 2013

Simester D (2011) When you shouldn't listen to your critics. Harvard Business Review June 2011, 42

Sloane G (2012) Groupon hurt by lack of repeat biz. New York Post. http://www.nypost.com/p/news/business/daily_deal_downer_BD1cnhAMIIINGF8bU6qOeP. 4 Jan 2012, Accessed 15 June 2013

Sonnenschein M (2011) Getting the best out of online reservations. Gourmet Marketing. http://www.gourmetmarketing.net/2011/04/11/getting-the-best-out-of-online-reservations. 11 April 2011, Accessed 15 June 2013

Stein JD (2010) Graham Elliot Bowles on haters, pills, and being on fox. Eater. http://eater.com/archives/2010/07/27/graham-elliot-bowles-on-haters-pills-and-being-on-fox.php. 27 July 2010, Accessed 15 June 2013

Steiner C (2010) Meet the fastest growing company ever. Forbes.com. http://www.forbes.com/forbes/2010/0830/entrepreneurs-groupon-facebook-twitter-next-web-phenom.html. 12 Aug 2010, Accessed 15 June 2013

Stoppelman J (2010) Yelp by the numbers, Yelp web log. http://officialblog.yelp.com/2010/12/2010-yelp-by-the-numbers.html. 15 Dec 2010, Accessed 15 June 2013

Streitfeld D (2011a) In a race to out-rave, 5-star web reviews go for $5. New York Times. http://www.nytimes.com/2011/08/20/technology/finding-fake-reviews-online.html. 19 Aug 2011, Accessed 15 June 2013

Streitfeld D (2011b) Coupon sites are a great deal, but not always to merchants. New York Times. http://www.nytimes.com/2011/10/02/business/deal-sites-have-fading-allure-for-merchants.html. 1 Oct 2011, Accessed 15 June 2013

Stross R (2008) How many reviewers should be in the kitchen? New York Times. http://www.nytimes.com/2008/09/07/technology/07digi.html. 6 Sept 2008, Accessed 15 June 2013

Sutel S (2007) For latest reviews, chefs look online. Washington Post. http://www.washingtonpost.com/wp-dyn/content/article/2007/07/03/AR2007070301345_pf.html. 3 July 2007, Accessed 15 June 2013

Tan A, Kucera D (2012) Investors give Yelp a rare review as shares soar in market debut. Sydney Morning Herald. http://www.smh.com.au/business/investors-give-yelp-a-rave-review-as-shares-soar-in-market-debut-20120303-1u9p2.html. 4 March 2012, Accessed 15 June 2013

Tedeschi B (2004) Click and eat on the road. New York Times. http://www.nytimes.com/2004/05/23/travel/practical-traveler-click-and-eat-on-the-road.html?pagewanted=all&src=pm. 23 May 2004, Accessed 15 June 2013

Trapunski R (2011) How the Internet has changed Toronto's dining scene. Toronto Globe and Mail. 21 May 2011

Weingarten E (2010) Forget journalism school and enroll in Groupon academy. Atlantic. http://www.theatlantic.com/technology/archive/2010/12/forget-journalism-school-and-enroll-in-groupon-academy/68257/. 20 Dec 2010, Accessed 15 June 2013

Weiss B (2010) Groupon's $6 billion gambler. Wall Street Journal. http://online.wsj.com/article/SB10001424052748704828104576021481410635432.html. 20 Dec 2010, Accessed 15 June 2013

Wilkes D (2011) Cupcake calamity: Groupon discount deal leaves baker swamped by orders for 102,000 cakes and wipes out her profits. Daily Mail. http://www.dailymail.co.uk/femail/article-2064208/Cupcake-calamity-Website-discount-deal-leaves-baker-swamped-orders-102-000-cakes-wipes-profits.html. 23 Nov 2011, Accessed 15 June 2013

Willett M (2012) Horrible Yelp reviews of New York's Michelin-starred restaurants. Business Insider. http://www.businessinsider.com/yelp-reviews-of-michelin-restaurants-2012-10?op=1. 9 Oct 2012, Accessed 15 June 2013

Williams G (2010) Groupon's Andrew Mason: The unlikely dealmaker. AOL Small Business. http://smallbusiness.aol.com/2010/08/09/groupons-andrew-mason-the-unlikely-dealmaker/1#c29811638. 9 Aug 2010, Accessed 15 June 2013

Wong W (2012) Groupon stock falls to all-time low as IPO anniversary nears. Chicago Tribune. http://articles.chicagotribune.com/2012-11-02/business/chi-groupon-stock-falls-to-alltime-low-as-ipo-anniversary-nears-20121102_1_groupon-stock-chief-executive-andrew-mason-groupon-getaways. 2 Nov 2012, Accessed 15 June 2013

Yiannopoulos M (2011) Groupon's image problems spread to Europe. Telegraph. http://www.tele-graph.co.uk/technology/social-media/8413336/Groupons-image-problems-spread-to-Europe.html. 29 March 2011, Accessed 15 June 2013

Yoo A (2009) Nasty altercation between Yelp critic, bookstore owner. The Scavenger (blog), San Francisco Chronicle. http://blog.sfgate.com/scavenger/2009/11/05/nasty-altercation-between-yelp-critic-bookstore-owner. 5 Nov 2009, Accessed 15 June 2013

Yu R (2011) Social-media sites give travelers local insights. USA Today. 1http://travel.usatoday.com/digitaltraveler/2011-02-11-digital-traveler_N.htm. 15 Feb 2011, Accessed 15 June 2013

Zetter K (2010) Yelp Accused of Extortion. Wired. http://www.wired.com/threatlevel/2010/02/yelp-sued-for-alleged-extortion. 24 Feb 2010, Accessed 15 June 2013

Chapter 4
Trust Online: From E-Commerce to Recipe Sharing

Trusting is hard. Knowing who to trust, even harder.

(Snyder 2008 as quoted in goodreads.com)

The previous chapter described the untrustworthiness of certain Yelp reviews, for example reviews written by family or friends that are unfairly favorable or those written by competitors or dismissed former employees that are unfairly unfavorable. While Yelp uses both a computer-based algorithm and a backup human review process to remove unfair reviews, they are not able to filter out all unfair reviews by their processes.

This chapter continues the discussion of trust online. The first section briefly considers some of the earliest examples of unfair online community reviewing. But the majority of the chapter is focused on the trustworthiness of recipes that are posted and shared online. The chapter presents six models used to make readers trust recipes that are posted online. The final section of the chapter looks at the information studies literature on trust, and how that literature addresses issues of trust in both restaurant reviewing and recipe sharing.

4.1 Early Community Reviewing Online

Online community reviewing was pioneered by the e-commerce giant Amazon, and many of the early instances of biased reviews are associated with reviews for books, computer equipment, and other items posted on the Amazon website. Figure 4.1 presents some of the most prominent early examples appearing on Amazon and elsewhere. These examples are merely indicative of the problem, not a comprehensive listing.

Many companies that rely on a crowdsourcing reviewing process are eager to find ways to reduce or eliminate spurious reviews. Amazon, like Yelp, has a ranking system for reviewers. Some websites require self-declaration of conflict of interest,

W. Aspray et al., *Food in the Internet Age*, SpringerBriefs in Food,
Health, and Nutrition, DOI 10.1007/978-3-319-01598-9_4,
© William Aspray, George Royer, Melissa G. Ocepek 2013

- Novelist Lev Grossman wrote an article on the news and entertainment website Salon explaining how he posted anonymous positive reviews on Amazon of his own work after his first novel, *Warp*, which was published in 1997, received negative reviews there.
- In 2009, Elsevier, the publisher of a textbook on *Clinical Psychology*, offered all contributors a $25 Amazon gift certificate if they would write favorable book reviews on the Amazon and the Barnes and Noble websites. When reported in the press, Elsevier's director of corporate relations made a public statement that this was an employee error and that no gift certificates were given out for this purpose.
- In 2009, it became publicly known that a business development representative of Belkin had been paying people 65 cents to write a 5-star (most favorable) review on the websites Amazon, Newegg, or Buy of one of the company's products – a router that had previously received numerous negative reviews for being buggy, unreliable, and expensive. Reviewers were instructed to write their reviews without having ever seen the product, and also to mark any negative reviews they found as "not helpful." When confronted with the news, the Belkin president indicated that he was not aware and did not sanction these activities of this employee.
- In 2010 Orlando Figes, a professor of Russian history at Birkbeck College, University of London, posted several reviews on the UK branch of Amazon – criticizing the work of two other historians of Russia (Robert Service and Rachel Polonsky) and praising his own work. Figes eventually paid legal costs and damages and issued a public apology to Service and Polonsky.
- Reporter David Streitfeld of the *New York Times* reported in 2011 on an offer on the help-for-hire website Fiverr: "For $5, I will submit two great reviews for your business"; he also reported at the same time on a work-for-hire offer on *Digital Forum*: "I will pay for positive feedback on TripAdvisor."
- It was also reported in 2011 by a University of California Santa Barbara computer science professor that two Chinese crowdsourcing websites, Zhubajie and Sandaha, are 88 and 92%, respectively, devoted to "crowdturfing", a term coined to refer to using crowdsourcing methods to influence public opinion with fake activities (a play on the term "astroturfing", which refers to influencing public opinion with fake grassroots activity). A typical fake placement costs much less than a dollar, and Zhubajie is reportedly paying out more than a million dollars a month for crowdsourcing tasks. The same article also claims that ShortTask, the second largest US crowdsourcing website after Amazon's Mechanical Turk, devotes 95% of its efforts to crowdturfing tasks.
- In 2012 it was reported that VIP Deals, a supplier of leather cases for Amazon's Kindle Fire tablet computers, gave customers the cost of the case ($35) for writing a favorable review on Amazon. This brought a statement of concern from the Federal Trade Commission.

Fig. 4.1 Examples of biased online reviews (Source: Back 2010; Baez 2009; Chapman 2011; Domenic 2012; Grossman 1999; Humphries 2009; Lewis 2011; McCracken 2009; Ott et al. 2011; Parsa 2009; Simonite 2011; Soulskill 2009; Streitfeld 2011; Wang et al. 2012)

e.g. if the person writing the review is an employee of the company that makes the product; however, there is no compliance mechanism and only weak sanctions (Lamm 2012; Lubin 2012; Puget Sound Business Journal 2012; Silberman et al. 2010). Some industry analysts believe that the best long-term solution is to rely on reviews written by people one knows, e.g. people who are one's Facebook friends. There are firms – Main Street Hub being a notable example – that manage the online reputations of small businesses for a fixed fee (reportedly for about $200 per month) (Higgins 2011).

- The tone does not fit.
- The reviewer's profile is so incomplete as to make the reviewer anonymous.
- The review is all good or all bad.
- The review is out of date.
- The reviewer seems too familiar with the staff or suppliers to be able to be objective.

Fig. 4.2 Telltale signs of suspicious reviews (Source: Ott et al. 2011. Also see Leung 2010)

Researchers at Cornell University have devised an algorithm to detect fake reviews (what they called "deceptive opinion spam"), which in their test was 90 % accurate in distinguishing between 400 fake, positive reviews of Chicago hotels created using Amazon's crowdsourcing service Mechanical Turk when they were mixed with 400 authentic, positive reviews from the social networking review website TripAdvisor. Their study found that the fake reviews tended to use a large number of superlatives, were generally not strong on the descriptions of the hotels, focused more on why the reviewer was in Chicago than on the hotel, and used "I" and "me" more frequently. In Fig. 4.2, senior employees of Yelp, Chowhound, and OpenTable give some practical guidelines about how to spot a suspicious review.

4.2   Finding and Sharing Recipes

One of the most popular food activities online is to search for, share, adapt, and comment on recipes. For example, Allrecipes includes more than 40,000 recipes, has more than 3 million registered members, and receives 400 million visits per year (Furchgott 2010). But how can a user know to trust a recipe that she – or, increasingly, he – finds? This section discusses six websites, each of which represents a different approach to providing trusted recipes: the community review model of Allrecipes, the expert chef model of ten chefs who are popular online, the laboratory testing model of *Cook's Illustrated*, the scientific research method of Modernist Cuisine, the corporate publishing model of Conde Nast, and the corporate food products model of General Mills.

4.2.1   The Community Review Model: Allrecipes

In 1997, 2 years after completing his graduate study in archeology at the University of Washington, Tim Hunt was working as a lead developer (later CEO and president) of Emergent Media, a web development company. After searching fruitlessly online for a favorite cookie recipe, he founded cookierecipes, for everyday bakers to share their personal recipes online. The website was popular, and Hunt then formed BreadRecipes and other websites for other specific kinds of food recipes. Eventually these were merged into Allrecipes. Hunt hired Bill Moore, an executive at Starbucks who had invented the Frapuccino, to run the company. By 2006, when Allrecipes was purchased by the publisher Reader's Digest Association for $66 million, it had 1.8 million registered members (Allison 2006). Reader's Digest already published two leading food magazines, *Taste of Home* and *Every Day with Rachel Ray*, and it believed that Allrecipes would serve as a good portal to its various food magazines, cookbooks, and websites. The website continued to grow under the direction of Reader's Digest, until it was purchased in 2012 for $175 million by Meredith Corporation, the leading media and marketing publisher for American women. Meredith's publications include *Better Homes and Gardens*, *Parents*, *Family Circle*, and *Ladies' Home Journal*, among others (Groves 2010; Meredith 2012; Reader's Digest Association 2012).

Recipes on the Allrecipes website are overwhelmingly designed for foods that ordinary people make at home, not fancy chef-inspired creations. Sixty-four percent of the website visitors are women between the ages of 25 and 55. The website supports four kinds of searches. There is a general search, e.g. put in the word "cupcake" and find all recipes for either making cupcakes or that use cupcakes as an ingredient. A second search tool enables more complex searching by allowing one to add a list of ingredients to be used and a list of ingredients to be avoided. A third search tool focuses on nutrition, e.g. cupcakes sorted by the amount of calories, fat, or carbohydrates. Finally, an advanced search tool considers multiple factors at once: preparation time, cooking time, meal (e.g. breakfast), dish type (e.g. main dish or salad), cooking method, main ingredient, style (e.g. gourmet or kid pleaser), ingredients to include or exclude, dietary preference (e.g. gluten-free), and recipe sponsor (e.g. Hellman's or Betty Crocker).

Any user of Allrecipes – whether the person is a member or not – can search or browse for recipes. Browsing categories include food courses (e.g. salad or dessert), ingredients and methods (e.g. barbecuing and preparing pasta), and occasions and cooking styles (e.g. Thanksgiving or world cuisines). There are special tabs for menus (e.g. quick and easy or in-season), cooking tips, and how-to videos. In order to post blogs, add pictures, or write reviews, however, one must log in as a member. There is no cost to become a member. Members can rate any recipe appearing on the Allrecipes website, using a five-star system, and also write a review of any recipe. For each review, there is a profile of the reviewer, often giving the reviewer's photo, hometown, current city of residence, cooking interests, and a list of the reviewer's own recipes and other reviews. There is also a way for others to indicate whether they found a particular review helpful.

For example, if one searched on "chocolate cake," one would find 1,275 recipes (as of March 15, 2012). One of the recipes near the top of the list, called Too Much Chocolate Cake, received a 4.76 star rating based on the scores of more than 4,500 raters. The recipe also had more than 3,500 reviews. Reviewer comments ranged from how well they liked the recipe, to stories about how they had made it for their special someone on Valentine's Day, to how to tinker with the ingredients or cooking time or temperature.

Allrecipes is available not only on one's computer but also on all the major mobile phone platforms. One of the amusing apps for the cellphone is Dinner Spinner: if one spins separate dials for dish type, ingredients, and preparation time, the app comes up with dishes from the Allrecipes database that meet that configuration, if there are any. The authors' first spin to test out this app resulted in Fish, Cookies, and Under 45 min, but there were no qualifying recipes. A second spin resulted in Breakfast, Fruit, and 20 min or Less. One hundred and eighty recipes fit these criteria, including numerous fruit smoothies, apple-cinnamon oatmeal, whole-wheat blueberry pancakes, and grilled peanut butter and banana sandwiches, among others.

Trust of the recipes on this website comes from the fact that these are recipes for everyday folks by everyday folks (not chef creations), that many of the recipes have large numbers of ratings and reviews, that others label individual reviews of a recipe as helpful or not, and that there is some information about each reviewer in order to help reach one's own conclusions as to whether to trust a particular review. Moreover, the prevailing use of these recipes seems to be for everyday cooking; thus it might not be such a big deal if a particular recipe is less than satisfactory than it might be if one were cooking for a major occasion.

4.2.2   The Laboratory Testing Model: Cook's Illustrated

In 1980, at age 29, Christopher Kimball launched *Cook's* magazine in the garage of his home in Weston, Connecticut. At the time, Kimball had limited experience with either publishing or cooking. While he was growing up, his college professor mother grew organic vegetables but did not cook; the family employed someone to cook the family meals. Kimball first learned to cook in the summers of his childhood from a family friend on the family farm in Vermont. After graduating from Philips Exeter Academy and Columbia University, where he received an undergraduate degree in primitive art, he took a job in his stepbrother's small publishing firm and later worked for the Center for Direct Marketing, an organization that ran courses on marketing and publishing. While working at the center, he began taking cooking classes but was deeply dissatisfied with his instructors' answers to his numerous questions about why to do things in the way they instructed. He thus decided to start his own magazine on home cooking, using $110,000 in startup funds provided by family and friends.

There are not many good job opportunities for food writers, and this helped Kimball to assemble a talented staff. Several of his editors went on to become leading cookbook authors. After struggling for 3 years to attract enough advertising and deal with the other business aspects of running the magazine, he sold partial interest in his company to the *New Yorker* magazine. In turn the magazine was sold to Advance Publications (Conde Nast), then again to the Swedish publisher Bonnier Group, and finally back to Advance, which ceased its publication in 1989. Kimball departed for a career in publishing: first starting a men's magazine that failed before the first issue appeared, then taking over a macrobiotic magazine named *East West Journal*, which he took mainstream and retitled *Natural Health*. Selling *Natural Health* for more than $15 million, he now had funds to return to the cooking publication business.

Kimball learned that Advance Publications had failed to protect the trademark for the name Cook's, so he bought it for $175, rehired several of the editors from his former cooking magazine, and started his new magazine *Cook's Illustrated* in 1993. In words that may be Kimball's own, "It's a magazine for people interested in understanding the techniques and principles of good home cooking. Each article dissects well-known cooking methods and ingredients with one goal in mind: to develop the simplest, most foolproof recipes for the best-tasting result" (PR Newswire 2007; Sagon 2004; Stohs 2009). It must not have seemed to outsiders like a very promising business plan: there was no color inside the magazine (only on the cover), no ads, the magazine appeared infrequently (only every other month), there was no discounting of subscriptions, and it was expensive at the cost of $25 for six 32-page issues a year. Nevertheless, the magazine's circulation grew rapidly, from 25,000 in its first year to more than a million by 2007. The magazine had a renewal rate of almost 80 %, twice that of most consumer magazines. It attracted a surprising number of men – 40 % of subscribers – who possibly were attracted by the scientific approach of the articles (Matus 2009; PR Newswire 2007).

This venture expanded into a media empire including another magazine on home cooking that includes reader submissions, *Cook's Country*, which is a direct competitor to Reader's Digest Association's *Taste of Home* (the best-selling food magazine, with a circulation over five million). The empire also includes two nationally syndicated television shows on public broadcasting, *America's Test Kitchen* (with more than three million viewers) and *Cook's Country TV* (with more than a million viewers). There is a book publishing arm, which has published *The Best Recipe* and *America's Test Kitchen Family Cookbook*, as well as a series of more specialized cookbooks on topics such as baking or cooking steaks and roasts. There is also a website, which requires a paid subscription separate from the subscription to the magazine. With more than 150,000 paid subscribers to the website, this is the fastest growing part of the business today.

Kimball explains the philosophy behind the articles in *Cook's Illustrated*: "We develop recipes defensively...We not only test all the obvious ways to make something, but we also try to figure out what people will do to a recipe to cause it to fail, [such as] bad substitutions, the wrong oven temperature, the wrong flour, cheap cookware" (Kimball as quoted in Sagon 2004). The feature recipe in each issue is presented as a narrative, beginning with the reason the staff is making the recipe (e.g. they remember it fondly from their youth), following through the various failed efforts to get the recipe correct in their test kitchens, and ending up with what they argue is a foolproof, best recipe for this particular dish. The editors have found that readers like hearing about kitchen disasters, so they liberally describe the failures along the way to their "perfect" recipe.

The process for preparing a recipe for publication is painstaking. Each recipe is pre-surveyed to see if there will be reader interest. For example, readers have told the editors they are not particularly interested in hearing about candy or any seafood other than salmon and shrimp. A literature review is then conducted in the company's cookbook library to identify recipes related to the item to be cooked. These recipes and variations are cooked in the laboratory kitchens and tasted by the staff.

The staff offers critiques on ingredients and cooking methods, and the process is iterated, sometimes as many as 40 times. When the entire staff believes the recipe is as good as it can be, it is sent outside to a small group of trusted readers who make it in their own kitchens and send back reports. If the recipe is not well liked by these testers, it is scrapped. Surviving recipes are published, together with the story of their journey through the test kitchens. There is a science consultant to conduct experiments and add scientific explanations to the text. The recipes only use ingredients that can be found in any supermarket. The goal is consistency and predictability as well as excellent taste.

Each issue also includes some other recipes (typically around ten), which receive more abbreviated treatment; a test of various brands of some ingredient, such as balsamic vinegar, and reports on the use of that ingredient in various recipes; a test of some kitchen utensil, such as paring knives; a section offering how-to tips; updates on variations of recipes that have appeared previously in the magazine; and staff replies to reader questions.

Kimball's stated goal is to duplicate the reality and perception of fair and impartial testing of recipes, ingredients, and cooking equipment that *Consumer Reports* has achieved for household goods. The use of everyday ingredients, the attempt to produce home cooking, and the repeated laboratory testing all contribute to make people trust these recipes. Kimball stopped reviewing cookbooks in his magazines when his company started to publish its own cookbooks, to ensure this sense of fairness and impartiality. Kimball also stopped the use of freelance writers when it became clear that they could not uphold the rigorous standards in their tests of secondary recipes, ingredients, and cooking utensils. Complaints raised about Kimball include questions about whether there is, as Kimball believes, a single best version of a recipe; also whether this kind of cooking by consensus delivers watered-down taste. No one questions whether these recipes are trustworthy or practical.

4.2.3 The Scientific Model: Nathan Myhrvold and Modernist Cuisine

In 2011, Nathan Myhrvold, together with Chris Young and Maxime Bilet, published *Modernist Cuisine: The Art and Science of Cooking*. Tim Zagat called it "the most important book in the culinary arts since Escoffier." Noted New York chef David Chang called it "the cookbook to end all cookbooks." Mark Gemein, in *The Paris Review*, compared it to Jean Anthelme Brillat-Savarin's *Physiology of Taste*, written in 1825 and in print continuously since then – perhaps the most important book ever written on food (Brillat-Savarin 1825/2011). The $625, 6-volume, 2000-plus page treatise was the result of a 4-year effort by 46 chefs and assistants in Myrhvold's kitchen laboratory at his high-tech intellectual property firm Intellectual Ventures (Huffington Post 2011; Kummer 2011; Mullenweg 2011; Park 2011; Ruhlman 2011).

The 1,500 recipes in these books cover every area of cooking except pastry. These are not books for the home cook to use in everyday cooking, not even for that special dinner they are planning to make. These recipes require numerous specialized tools, many of them adopted from the processed food industry, that the ordinary cook does not have access to and could not afford to buy (immersion circulators, centrifuges, ultrasonic cleaning baths, rotary evaporators) as well as high-end but more common kitchen tools such as pressure cookers, food sealers, and grinders. The recipes are exceedingly meticulous, complicated, and time consuming – the hamburger recipe, for example, requires 30 h of preparation time (Gunnison 2011). Even most professional chefs will not cook directly from most of the recipes, but the books are useful to teach techniques as well as to educate about the scientific principles behind many of the chemical and physical processes that take place during cooking.

There is also a website, Modernist Cuisine: The Art and Science of Cooking. It advertises the six volumes, lists events such as book signings, and provides a recipe library, a cooking forum, a guide to gear for the modernist kitchen, a place to shop for memorabilia such as tee shirts with an image of the 30 h hamburger silk screened on it, and a blog discussing such topics as how to prepare the perfect turkey for Thanksgiving. One recipe, for caramelized carrot soup, announces in big letters that no centrifuge is necessary to make this recipe!

These books can be seen as the apotheosis of a food movement sometimes known as the molecular gastronomy movement. Four highly acclaimed chefs – Ferran Adria in Spain, Heston Blumenthal in England, and Thomas Keller and Grant Achatz in the United States – have been the intellectual leaders of this movement (Frieswick 2009). Adria is the head chef at El Bullia restaurant in the Catalonia region of Spain. *Restaurant* magazine has pronounced his as the best restaurant in the world on four occasions. Adria prefers to call his cooking style "deconstructivist." In the late 1980s he began doing experiments in his restaurant using science to understand culinary practice. He closed the restaurant in 2012 for 2 years, to give him time to reflect on the future direction of his and his restaurant's cooking (Moore n.d.; Toomey 2010).

Blumenthal is the owner of The Fat Duck, a restaurant in Bray, England that once won the top honor from *Restaurant* and several times came in second to El Bullia (Blumenthal 2010). Blumenthal studies the molecular composition of dishes to better understand their taste and flavor. In his book *In Search of Perfection*, he traveled the world in his quest to make the perfect version of 16 classic dishes. For example, he travelled to Delhi, India to take an MRI of a chicken as part of his effort to make the perfect Chicken Tikka Masala.

Thomas Keller is the chef and owner of two well-known restaurants, The French Laundry in California and Per Se in New York, both of which have been awarded three stars in the Michelin Guide. He is also author of several award-winning cookbooks.

Achatz is the owner of a Chicago restaurant Alinea, which won the *Gourmet* magazine award for best restaurant in America in 2006. In 2008 he won the James Beard Foundation award as Best Chef in the United States (McClusky 2006). He was trained at the Culinary Institute of America and then served under Thomas Keller as sous chef at The French Laundry. Achatz is an extensive user of esoteric

- "food at its most ridiculous"
- "culinary Mount Everest"
- "This is just like anything else, there is a point of diminishing returns for the average person."
- "I think a lot of people are looking at this cookbook the wrong way. It's not 'this is how you should make it' type of thing. It's an aggregation of all the food knowledge we have to this point, and really a triumph in technique."
- "Yes, this burger is ridiculous on some level, but that's missing the point. This is the food equivalent of a concept car. Most concept cars would be absurd on the road, but they aren't meant to be market-ready vehicles. They are meant to convey ideas."
- "As a scientist, cook, and avid eater, I find this book to be fascinating."
- "*Modernist Cuisine* is like the food analog of Euclid's *Elements*."

Fig. 4.3 Reviews of *Modernist Cuisine* (Source: Lopez-Alt 2011)

laboratory equipment such as antigriddles (to flash-cool foods to −30°), defusers, dehydrators, induction burners, and an immersion circulator. Not content with the laboratory equipment supply companies have to offer, he has entered into collaboration with blacksmith and sculptor Martin Kastner to create new kitchen tools that are at once beautiful and functional (Kastner 2011).

The publication of Myhrvold's book has created extensive discussion in online forums, with a wide range of viewpoints expressed. Figure 4.3 presents a sample of these views. While the books may not be practical, there is strong trust in the results. One commentator called it "as scientific as it is gastronomic" (Ruhlman 2011). Another commentator points to the authors as being "detail-oriented and [having] science-driven obsession with quality" (Mullenweg 2011). Recipes are laboratory precise, often measuring ingredients to 1/100th of a gram. Chemical and physical processes are described in great detail. The equipment used is the best that money can buy.

Modernist Cuisine has a further cachet because Myhrvold himself is such an intriguing and controversial figure. He started college at age 14 at UCLA, where he studied mathematics and physics. He earned a masters degree in mathematical economics and a Ph.D. in physics from Princeton. He then worked on cosmology as a postdoc with Stephen Hawking at Cambridge. After his education was complete, he co-founded Dynamical Systems Research, a software firm, which was eventually acquired by Microsoft. During his 13 years at Microsoft, he was awarded numerous patents and founded Microsoft Research. He left the company in 2000 to co-found Intellectual Ventures, a technology and energy patent acquisition firm that is regarded by some as a patent troll. He is a master chef. He spent between one and ten million dollars of his personal funds – he won't specify how much – to pay for the laboratory and people that resulted in these books.

While many people question whether the recipes are worth all the extra time and expense for a marginal improvement in the taste, there are a few critics who simply believe the food that results is not very good. One outspoken critic is John Mariani, a food and wine correspondent for *Esquire*: "This is the sort of thing that's being done at General Foods and CPC and Purina...That's what these chefs are trying to pass off as cuisine, when often it's just not very tasty or tasteful" (McClusky 2006; for context, also see Mariani 2012).

While Myhrvold might go to extremes to attain the sublime, there are others who have popularized the science of food and shown the general public how scientific principles enable the everyday cook to gain better results. The earliest notable example is Harold McGee's book, *On Food and Cooking: The Science and Lore of the Kitchen*, which first appeared in 1984 and was revised and expanded for its twentieth anniversary (McGee 2004). A more recent contribution in the same vein is Jeff Potter's *Cooking for Geeks: Real Science, Great Hacks, and Good Food* (2010). Perhaps the best known of these efforts in recent years has been by Alton Brown, whose television show *Good Eats* ran on public television and later The Food Network from 1998 to 2011. Brown has also published about a dozen books that are spinoffs of his television show. These authors have the authority of science behind their recipes.

4.2.4   The Expert Chef Model: Ten Famous Chefs

In 2009, Tony Mamone, the CEO of online publisher Zimbio, identified what he calls the 10 Most Famous Chefs in the world by counting the number of online searches on their names. Mamone's list is given in Fig. 4.4 and our analysis of these chefs is given in Fig. 4.5.

What is clear from Fig. 4.5 is that, in order to be on Mamone's list, media presence is more important than being a thought leader in the cooking field; and having been professionally trained in a high-quality culinary school or having gained experience as a chef or sous chef at a high-end restaurant is not necessary. For example, Sandra Lee, Jamie Oliver, and Rachel Ray have little or no culinary training or high-end restaurant experience. One might argue that the thought leader in the international culinary arts scene does not appear on Mamone's list. A strong case can be made to award that position to Joel Robuchon, with his recognition by Gault Millau as the "Chef of the Century," the 26 Michelin stars attained by his various restaurants, his leadership role in *Larousse Gastronomique*, and numerous other awards. One might argue that in America Julia Child was enormously influential in bringing French cooking back to America through her co-authored cookbook *Mastering the*

Jamie Oliver –The Naked Chef
Gordon Ramsay – Hell's Kitchen
Rachel Ray – Food Network Queen
Bobby Flay – BBQ Throw Down
Wolfgang Puck – Restaurant Tycoon
Giada de Laurentiis – Everyday Italian
Sandra Lee – Semi-Homemade Cooking Author
Mario Batali – Iron Chef Champion
Emeril Lagasse –"Kick It Up a Notch"
Jacques Pepin – Famed French Chef

Fig. 4.4 Tony Mamone's list of the ten most famous chefs in the world (Source: Mamone 2009)

Name	Bio	Recipes	Foundation	Books	TV	Restaurants	Shop	Video	Skills	Blog/Articles	Other
Oliver	x	+++	x	x	x	x	x	x	x		
Ramsay	x				x	x					
Ray	x	+++	x							x	Kids, pets
Flay	x	++		x			x		x		
Puck	x	++				x	x				Catering, reservations, company info, franchising
De Laurentiis	x	++			x		x	x	x		
Lee	x	+	x	x				x			Recipe sharing
Batali	x	+	x	x		x	x				travel
Lagasse	x	+++	x			x	x	x	x	x	
Pepin	x	+		x	x			x			TV behind the scenes

Fig. 4.5 Content analysis of websites of Mamone's ten most famous chefs

Art of French Cooking and her long-running television show *The French Chef*. Or one might choose Alice Walters for her role in introducing organic food and local food into America through her restaurant Chez Panisse and her cookbooks. Julia Child, who died in 2004, does not have a website. Joel Robuchon's and Alice Water's websites are more about their restaurants than about providing recipes or cooking tips. Despite all these limitations, we use Mamone's list for analysis below.

All ten people on Mamone's list include much more on their websites than recipes. The websites are more generally about selling the person and the person's business interests than anything else. Every one of the ten has significant biographical information on the website, often with information not only on his background but also on upcoming events at which he will appear. Seven of them include e-commerce pages on which they sell their branded products, such as food products or kitchen implements. Five have information about the (often multiple) restaurants that they own, and in the case of Wolfgang Puck there is even information about franchising opportunities for his eponymous restaurants. Six have videos or give tips in some other fashion, about preparing particular recipes or employing particular cooking techniques. Five include information on their website about their television shows. Four have information about cookbooks or other books, such as autobiographies, they have written. Four have information about their foundations or about the charities they support. Rachel Ray and Emeril Lagasse include blogs or reprint articles they have written elsewhere. Rachel Ray has a tab for moms about feeding their kids and another tab for dealing with pets including pet rescues and pet food recipes.

The websites vary considerably in the number of recipes they offer. Jamie Oliver, Rachel Ray, and Emeril Lagasse have the most (more than 1,000) recipes on their websites. Bobby Flay, Wolfgang Puck, and Giada de Laurentiis include a moderate number of recipes (from 100 to a few hundred). Sandra Lee and Jacques Pepin include a few recipes (under 100). Gordon Ramsay does not include recipes on his website.

People may trust these recipes because these individuals are familiar from having been well-known television personalities. Alternatively, it may be that readers believe that these people are engaged in sufficiently large business enterprises that they have talented people to test the recipes so as to avoid any unfortunate kitchen catastrophes that might blemish their public image. In a number of cases, however, trust does not reside in demonstrated kitchen talent through training or experience.

4.2.5   The Corporate Publishing Model: Conde Nast

Epicurious is a recipe website owned by the magazine publishing giant Conde Nast. The company publishes magazines for middle and upper income families, including *Bon Appetit, Gourmet, Architectural Digest, Bride's Magazine, House and Garden,* the *New Yorker, Vanity Fair,* and *Glamour.* Magazines generally have had difficulty in figuring out an effective strategy for using the web; most magazines have regarded it as a direct competitor, harming their magazine subscription numbers. In 2006, Conde Nast made several web acquisitions, including Wired News, Epicurious, Style, and Concierge. They were intended to be companion websites to the magazines that the company publishes on food, travel, and men's and women's style. The hope was that many of the more than three million monthly readers of the Epicurious website would sign up for Conde Nast print magazines (Hillebrand 1999; Terdiman 2006).

Epicurious includes many of the same features as other recipe websites such as recipes, blogs, and user reviews; but it also includes tools to create favorites lists, email recipes to friends, and prepare shopping lists. The website has extensive search functions that enable a reader to search by main ingredient, meal (breakfast, lunch, dinner), cuisine, diet (healthy, low fat, low calorie, low sodium, vegetarian, vegan, high fiber, kosher, low sugar, wheat/gluten free, raw, low carb), season, and holiday/special occasion (Judge 2009).

It is useful to compare Epicurious with Allrecipes. The latter contains a larger number of recipes (40,000 recipes compared to Epicurious's 28,000), has a flashier layout, and a more versatile search function. All of the recipes on Allrecipes are contributed by "home chefs," whereas the majority of recipes on Epicurious are drawn from *Gourmet* and *Bon Appetit* magazines, or cookbooks published by Conde Nast, or are provided directly by professional chefs. The person who is interested in more ordinary fare or who wants to participate in sharing recipes is more likely to go to Allrecipes, while the person interested in something special is more likely to go to Epicurious. Many people visit both websites regularly. Despite Allrecipes's many advantages, Epicurious has a strong following. As one food blogger said: "After using Epicurious, other food websites like Allrecipes and the odd blog just seem lackluster and uneventful by comparison, devoid of mouthwatering pictures and glowing reviews that seem to decorate every nook and cranny of Epicurious" (Judge 2009).

4.2.6   The Corporate Food Products Model: Betty Crocker

The back cover of the *Betty Crocker Cookbook: 1500 Recipes for the Way You Cook Today* touts itself as "America's Most Trusted Cookbook Made New." This book is the most recent edition of the popular cookbook first published in 1950, which has sold 65 million copies in its 11 editions. As the back cover goes on to claim, the cookbook offers "foolproof recipes, reliable how-to advice and delicious inspiration." Many people grew up with an earlier edition as the standard go-to cookbook in their home (although some families relied instead on *Better Homes and Garden Cookbook*, now in its 15th edition, having sold 40 million copies since its introduction in 1930; or *The Joy of Cooking*, introduced 1 year later, having gone through nine editions and chosen by the New York Public Library as one of the 150 most important and influential books of the twentieth century).

While *The Joy of Cooking* was created by a mother and daughter team, who tested the recipes in their Depression-era kitchen, the other two cookbooks are corporate products. What does it mean to trust a corporate product? Does it mean to trust the ethical behavior of the company? If you judge the food products that General Mills sells, such as Cheerios, Bisquick, Gold Medal Flour, and Nature Valley granola bars, to be of high quality, do you therefore trust the recipes in the cookbook? The fact that the company sells other products can readily be seen as having a negative impact on the objectivity of the cookbook. After all, the Betty Crocker cookbook evolved from a softcover recipe book printed by the company in the 1930s distributed as a means to sell more Bisquick. There never was a person named Betty Crocker involved with the company; she was a corporate brand name and image created by home economist Marjorie Husted in 1921 for the Washburn Crosby Company as a means to personalize responses to consumer questions. The name was retained when Washburn became a part of General Mills. Betty was chosen as the first name because it was believed to convey a cheery personality and was a common American name. The visual image of Betty Crocker that appears on various products and sales literature, which appeared first in 1936, has been updated 13 times and is intended to convey the image of a knowledgeable and caring homemaker. The current version is a composite of photographs of 75 real-life women of various ages and ethnicities. Should the fact that the company manipulates the corporate image of Betty make one doubt the cookbook? Does what Betty looks like really matter to the cookbook reader's trust? It may be that this mattered more in the 1930s than it does today, since many people today have a jaded view and expect corporate manipulation of the media. Of course, the updated images of Betty have to do with staying in fashion and in attracting a new generation of readers. Much of the trust has to do with remembering it was the go-to cookbook when you were growing up.

4.3 Information Studies About Trust

Trust has been studied by a number of academic disciplines, including organization theory, philosophy, political science, psychology, sociology, and transaction economics (He et al 2009; Hilligoss and Rieh 2008; Kim and Han 2009). The approach varies by field. For example, psychologists and political scientists treat trust as a psychological trait of the individual, whereas sociologists focus on the relationships across groups. The study of trust online has mostly been the province of information scholars, who draw not only on the literatures of the social and behavioral scientists, but also on research from their own subfields of information quality and human-computer interaction.

As more everyday activities are conducted online (for example, through the use of e-commerce or social question-and-answer websites), questions arise about whether a user can trust a particular system. For example, does a website such as PatientsLikeMe protect the privacy of the very sensitive health information the users have shared online, or in the case of an e-commerce website such as eBay does the company adequately protect the buyers' credit card information? The use of online systems also raises questions as to whether a user can trust the other users of the system. For example, why should a user trust someone else's review of a restaurant on Yelp or trust a recipe provided on Allrecipes, given that the reviewer is not an expert and is typically not known by the user? Empirical studies show that people are aware of their uncertainty and vulnerability when judging the quality of information that they find online.

The information studies literature addresses this latter question, about the trust one can have of the other people on the system, as well as the closely related question of the quality of the information other users provide. This literature makes a clear distinction – it is concerned with trust, which is a property of the information user and concerns dependability; it is not concerned with credibility, which is a property of the information itself and concerns its believability. Trust involves taking a risk, of having to suffer the consequences of acting on the information provided by others, e.g. a bad restaurant meal if one follows an untrustworthy review on Yelp or a failed cake if one follows an untrustworthy recipe on Allrecipes. Because users are different, it is not unreasonable to expect that they would reach different judgments about the trustworthiness of a particular piece of information they find online.

Many factors go into a decision to trust a piece of information gathered online. A user may consider the source of the information – is this restaurant review written by someone who is similar to me in valuing a white tablecloth dining experience, or is the reviewer someone who is young and highly value conscious? The user will pay attention to factual errors – a missing ingredient or too high a heat in a recipe. A user might also consider surface features, e.g. does the restaurant review seem too favorable as though it was written by a friend of the owner, or too unfavorable as though it was written by a competitor?

Alexander and Tate (1999) identify five criteria for assessing trust in information: accuracy and objectivity (both referred to in the previous paragraph), currency (is the information up to date about the restaurant?), coverage (does the review only talk about the furnishings and not the food?), and authority (is the person writing the review a trained chef or an experienced restaurateur?). Only the last of these five criteria – authority – is something that one might not be able to evaluate from the information itself.

Kelton et al. (2008) points out that there are a number of factors that go beyond the information itself that may cause a user to trust an online review. Here are several of Kelton's general points, expressed using food examples:

- Does the user have a predilection or aversion to using online reviewing systems such as Yelp?
- Has a friend recommended Yelp by telling how she uses it regularly to make decisions about where to eat?
- Is the user the type of person who has a high propensity to trust the recommendations of others about where to eat?
- Does the user have such confidence in her own cooking ability that she believes she can overcome any minor errors or unstated steps in cooking with a new recipe?
- How significant are the consequences of a bad restaurant review? For example, they might be minor if the user only has to travel across town to the restaurant and doesn't spend too much money, whereas the stakes are much higher if this is the only night to get a special meal in Paris while traveling.
- Has the user had a good experience with this particular website in the past?
- Has the user learned that the Betty Crocker website can generally be trusted for quick and simple meals for the family but is less trustworthy for romantic gourmet meals?

Other factors are also at play in the trust decision. You might be willing to trust a review from a particular restaurant reviewer because you have agreed with her reviews of other restaurants in the past. You might be willing to trust a Yelp reviewer because she has a large number of followers (Heffernan 2010). You might be willing to trust a restaurant review because a number of other people have published reviews, either on that website or some other website, saying approximately the same thing. You might trust a review because it is sensitive to issues that are important to you; for example, if you are elderly and the review addresses issues of how accessible and noisy the restaurant is, you might also be willing to trust its judgments about the food. You might trust a restaurant review because it appears on a website that others trust, e.g. the website has been written about favorably by a respected local chef or has been endorsed by an organization that aims for objective evaluation. You might trust a restaurant review because it is written with a verve, or flair, or crotchetiness that you can relate to. There might be particular words or a graphical design to the website where the review appears that makes it seem more trustworthy or more in line with your own personal interests (Flanagan and Metzger 2007; Fogg 2003; Fogg and Tseng 1999; Grabner-Krauter and Kaluscha 2003; Metzger 2007).

Some literature in information studies compares differences among novices (in the domain area), experts (in the domain area), and skilled information professionals (e.g. reference librarians or doctoral students who have a rigorous education in how to seek and evaluate information generically) (Brandgruwel 2005). Domain experts tend to pay more attention to factual accuracy than novices do. Lucassen and Schraagen (2011) argue that novices think about "surface level features" and use a "concrete line of argument", while experts are more likely to form abstract representations and focus on underlying principles. These authors have also found that novices are sometimes less trusting because they doubt their own abilities, so they wait to form trust until they have reasons to be highly confident of the information. It could be that this pattern is found less often in the realm of food because most people believe they have some experiential basis to talk about food, even if they are not an accomplished cook or gourmand.

In their model of information problem solving (problem definition, searching, scanning, processing, and organization of information), Brandgruwel et al. (2005) observe that experts place more effort than novices in the processing stage and scan information more frequently. There is not an exact correlate in restaurant or food reviews to the information professionals discussed in the information studies literature. Perhaps the closest parallel can be drawn between those heavy users of Yelp who, when reviewing a restaurant serving a cuisine unfamiliar to them, at least know the right questions to ask. A study by Golbeck and Fleischmann (2010) finds that textual cues in an online communication suggesting expertise or experience enhances trust among both expert and novices, whereas photographical images that give cues to expertise or experience enhance trust only among novices, not among experts.

The information studies literature discusses two types of trust – cognitive and affective. Cognitive trust is a belief that others will not act in an opportunistic manner but will follow through on their commitments, and assumes that there are protections in the system when someone misbehaves. Affective trust occurs when a user has an emotional bond to other users that enables her to feel secure and comfortable about relying on the system or a particular piece of information provided through it – even though there may not be adequate information to form cognitive trust (Sun 2010). While there are some features of Yelp, such as sequestering of objectionable and spurious reviews by Yelp staff and rating of reviewers by users, which might lead one to believe that cognitive trust is at play in Yelp, the depth of the trust in Yelp by active users is primarily a sign of affective trust.

4.4 Conclusions

The main section of this chapter has been focused on the trustworthiness of six types of websites for posting and sharing recipes, as well as for delivering other information related to food preparation. Each of these websites has strengths and weaknesses. The community review model is represented by Allrecipes, where any member can post a recipe for a particular dish and anybody – even without becoming a member – can read and download recipes. The website presents a problem of too many choices. One would not know which of 300 chocolate cake recipes found there would be good choices for one's own cooking, given that it is unlikely one would know anything about the person who had posted it, except for the ratings and comments from others online. In some cases, a recipe might receive thousands of reviews, some of which are bound to be contradictory.

The laboratory testing model is represented by *Cook's Illustrated*. The extensive testing of the recipe and its practical variations give the reader good reason to believe that the recipe is foolproof to prepare, or at least that all the most likely mistakes have been avoided. However, there is no reason to believe that there is one best recipe for any food item, or that the reader's taste will match that of the editorial staff of *Cook's Illustrated*.

The scientific model is represented by Nathan Myrhvold and the Modernist Cuisine movement. Their work helps one to understand the scientific principles behind cooking. However, there is nothing practical about their approach that can be translated into the home kitchen; the special equipment and elaborate procedures are far beyond anything that one would find in most homes or restaurants – so there is nothing practical to trust. Alton Brown provides a layman's version of scientific cooking, and this has attracted quite a following – including many young men who did not cook previously. Trust can be found in both the underlying scientific principles explained by Brown and in Brown's scientific explanation of the implications of some of the common mistakes made in the kitchen.

The expert model is represented by the websites and blogs of the ten professional chefs profiled above. We have seen that these websites are highly self-promotional, the basis for the chef's expertise is sometimes questionable, and celebrity often is conflated with expert knowledge of food and cooking.

The corporate publishing model is represented by Epicurious. The reason for trusting the recipes found there is that they have been prepared by expert chefs and have presumably been tested with care before being published in *Gourmet* or *Bon Appetit*.

The corporate food products model is represented by the Betty Crocker brand of General Mills. Trust resides in the familiarity with the cookbook, which has provided valuable service to everyday home cooks through multiple editions and many years, familiarity with the ingredients that are often household food product names produced by General Mills, and trust in Betty Crocker as a reliable and long-standing brand name.

Trust is an elusive pursuit. As the British screenwriter Troy Kennedy Martin (n.d.) once said, "I trust everyone. I just don't trust the devil inside them."

References

Alexander J E, Tate MA(1999) Web wisdom: How to evaluate and create information quality on the web. Erlbaum, Hillsdale, NJ

Allison (2006) The history of Allrecipes.com and the facts about Reader's Digest. Bfeedme (blog). http://www.bfeedme.com/the-history-of-allrecipescom-the-facts-about-readers-digest. 11 April 2006, Accessed 7 June 2013

Bäck EC (2010) Does Amazon Vine bias reviews? Internet & Technology (blog). http://elliottback. com/wp/does-amazon-vine-bias-reviews. 10 Dec 2010, Accessed 7 June 2013

Baez J (2009) Elsevier pays for favorable book reviews. The n-Category Café (blog). http://golem.ph.utexas.edu/category/2009/07/elsevier_pays_for_favorable_bo.html. 2 July 2009, Accessed 7 June 2013

Blumenthal H (2010) In search of total perfection. Bloomsbury. New York

Brandgruwel S, Wopereis I, Vermetten Y (2005) Information problem solving by experts and novices: Analysis of a complex cognitive skill. Comput in Hum Behav 21(3):487–508

Brillat-Savarin JA (1825/2011) The physiology of taste; Or, meditations on transcendental gastronomy. Reprint. Vintage, New York

Chapman N (2011) Fake online reviews and astroturfing. Atelier 186 (blog). http://www.atelier186.com/blog/2011/social-media/fake-online-reviews-and-astroturfing/. 20 Aug 2011, Accessed 7 June 2013

Domenic K (2012) Can readers trust paid-for book reviews? Guest blog post. http://lindadwelch.com/2012/01/can-readers-trust-paid-for-book-reviews. 31 Jan 2012, Accessed 7 June 2013

Flanagan AJ, Metzger MJ (2007) The role of site features, user attributes, and information verification behaviors on the perceived credibility of web-based information. New Media and Soc 9(2):319–342

Fogg BJ (2003) Prominence-interpretation theory: Explaining how people assess credibility online. In: Cockton G, Korhonen P (eds) Proceedings of the ACM CHI 2003 Conference on Human Factors in Computing Systems, ACM Press, New York, 722–723

Fogg BJ, Tseng H (1999) The elements of computer credibility. In: Proceedings of the Special Interest Group on Computer-Human Interaction (SIGCHI) at the Conference on Human Factors in Computing Systems, ACM Press, New York, 80–87

Frieswick K (2009) Perfection, Inc. Boston Globe Sunday Magazine. http://www.boston.com/bostonglobe/magazine/articles/2009/08/02/perfection_inc. 2 Aug 2009, Accessed 7 June 2013

Furchgott R (2010) Ready to start cooking? Fire up your iPad. New York Times. http://query.nytimes.com/gst/fullpage.html?res=9405EFDF163BF93BA25752C1A9669D8B63. 18 Nov 2010, Accessed 7 June 2013

Golbeck J, Fleischmann KR (2010) Trust in social Q&A: The impact of text and photo cues of expertise. Paper presented at the American Society of Information Science and Technology conference, Pittsburgh, PA, 22–27 Oct 2010

Grabner-Krauter S, Kaluscha EA (2003) Empirical research in on-line trust: A review and critical assessment. Int J of Hum-Comput Stud 58(6):783–812

Grossman L (1999) Terrors of the Amazon. Salon.com. http://www.salon.com/1999/03/02/feature_222. 2 March 1999, Accessed 7 June 2013

Groves S (2010) How Allrecipes.com became the worlds largest food/recipe site. ROI of Social Media (blog). http://www.stevengroves.com/2010/06/28/how-allrecipes-com-became-the-worlds-largest-food-recipe-site. 28 June 2010, Accessed 7 June 2013

Gunnison E (2011) Nathan 'The Terminator' Myhrvold takes on hamburgers. Eat Like a Man (blog), Esquire. http://www.esquire.com/blogs/food-for-men/nathan-myhrvold-burgers-060811. 8 June 2011, Accessed 7 June 2013

He W, Fang Y, Wei KK (2009) The role of trust in promoting organizational knowledge seeking using knowledge management systems: An empirical investigation. J of the Am Soc for Inf Sci and Technol 60(3):526–537

Heffernan V (2010) Accounting for taste. New York Times Magazine. http://www.nytimes.com/2010/10/10/magazine/10FOB-medium-t.html. 8 Oct 2010, Accessed 7 June 2013

Higgins M (2011) Crowd-sourcing for travel advice. New York Times. http://travel.nytimes.com/2011/08/21/travel/crowd-sourcing-for-travel-advice.html?pagewanted=all. 21 Aug 2011, Accessed 7 June 2013

Hillebrand M (1999) Williams-Sonoma, Epicurious make recipe for online sales. tech News World. http://www.technewsworld.com/story/537.html. 17 June 1999, Accessed 7 June 2013

Hilligoss B, Rieh S (2008) Developing a unified framework of credibility assessment: Construct, heuristics, and interaction in context. Inf Process and Manag 44(4):1467–1484

Huffington Post (2011) Modernist Cuisine has 'something for everybody' says Nathan Myhrvold. http://www.huffingtonpost.com/2011/03/24/nathan-myhrvold-modernist-cuisine_n_840373.html. 24 March 2011, Accessed 7 June 2013

Humphries M (2009) Belkin Is 'extremely sorry' after paying for good Amazon reviews. Geek. com. http://www.geek.com/articles/news/belkin-is-extremely-sorry-after-paying-for-good-amazon-reviews-20090120. 20 Jan 2009, Accessed 7 June 2013

Judge J (2009) Epicurious. AppStruck. http://appstruck.com/2009/07/iPhone-App-Review-epicurious. 8 July 2009, Accessed 7 June 2013

Kastner M (2011) Food informants: A week in the life of Martin Kastner, Alinea serviceware designer. Huffington Post. http://www.huffingtonpost.com/2011/07/14/food-informants-martin-kastner_n_895624.html#s308384&title=Antenna_Row. 14 July 2011, Accessed 7 June 2013

Kelton K, Fleischmann KR, Wallace WA (2008) Trust in digital information. J of the Am Soc of Inf Sci and Technol 59(3):363–374

Kim B, Han I (2009) The role of trust belief and its antecedents in a community-driven knowledge environment. J of the Am Soc of Inf Sci and Technol 60(5):1012–1026

Kummer C (2011) Better cooking through technology. MIT Technology Review. http://www.technologyreview.com/computing/37806. 21 July 2011, Accessed 7 June 2013

Lamm G (2012) Feds concerned about paying for Amazon.com five-star reviews. Tech Flash (blog), Puget Sound Business Journal. http://www.bizjournals.com/seattle/blog/tech-flash/2012/01/paying-for-amazoncom-five-star-reviews.html. 27 Jan 2012, Accessed 7 June 2013

Leung W (2010) Sniffing out fake restaurant reviews. Toronto Globe and Mail. 13 Oct 2010

Lewis R (2011) Beware rigged online reviews. CBS News. http://www.cbsnews.com/8301-502303_162-57318375/beware-rigged-online-reviews. 4 Nov 2011, Accessed 7 June 2013

Lopez-Alt JK (2011) Nathan Myrhvold's modernist burger. A Hamburger Today (blog), Serious Eats. http://aht.seriouseats.com/archives/2011/02/nathan-myhrvolds-modernist-burger.html. 3 Feb 2011, Accessed 7 June 2013

Lubin G (2012) The illegal way to improve your rating on Yelp. Business Insider. http://www.businessinsider.com/the-illegal-way-to-improve-your-rating-on-yelp-2012-2?utm_source=feedburner&utm_medium=feed&utm_campaign=Feed%3A+typepad%2Falleyinsider%2Fsilicon_alley_insider+%28Silicon+Alley+Insider%29. 19 Feb 2012, Accessed 7 June 2013

Lucassen T, Schraagen JM (2011) Factual accuracy and trust in information: The role of expertise. J of the Am Soc of Inf Sci and Technol 62(7):1232–1242

Mariani J (2012) Why it's hard to trust the Michelin standards. Eat Like a Man (blog), Esquire. http://www.esquire.com/blogs/food-for-men/michelin-guide-2013-13369832. 4 Oct 2012, Accessed 7 June 2013

Martin TK (nd) Quote by Troy Kennedy Martin: I trust everyone. I just don't trust the devil inside them. Goodreads. http://www.goodreads.com/quotes/63004-i-trust-everyone-i-just-don-t-trust-the-devil-inside. Accessed 10 May 2013

Matus V (2009) Cook's tour. Weekly Standard. http://www.weeklystandard.com/Content/Public/Articles/000/000/016/807ewdcz.asp?pg=2. 10 Aug 2009, Accessed 7 June 2013

McClusky M (2006) My compliments to the lab. Wired. http://www.wired.com/wired/archive/14.05/achatz.html?pg=1&topic=achatz&topic_set=. May 2006, Accessed 7 June 2013

McCracken H (2009) Is Belkin paying for fake favorable user reviews? Technologizer. http://technologizer.com/2009/01/17/is-belkin-paying-for-fake-favorable-user-reviews/. 17 Jan 2009, Accessed 10 March 2012

McGee H (2004) On food and cooking: The science and lore of the kitchen. Scribner, New York. Revised Edition

Mamone T (2009) 10 most famous chefs in the world. Zimbio.com. http://www.zimbio.com/Celebrity+Chefs/articles/d6Ex2fDmqkf/10+Most+Famous+Chefs+World. 14 Aug 2009, Accessed 7 June 2013

Meredith Corporation (2012) Meredith completes acquisition of Allrecipes.com from Reader's Digest. Press release. http://ir.meredith.com/releasedetail.cfm?releaseid=653206. 1 March 2012, Accessed 7 June 2013

Metzger MJ (2007) Making sense of credibility on the web: Models for evaluating online information and recommendations for future research. J of the Am Soc of Inf Sci and Technol 58(13):2078–2091

Moore B (nd) Ferran Adria. About.com: Gourmet Food. http://gourmetfood.about.com/od/chefbiographi2/p/ferranadriabio.htm. Accessed 17 March 2012

Mullenweg M (2011) Nathan Myhrvold and Modernist Cuisine. Matt: Unlucky at Cards (blog). http://ma.tt/2011/03/nathan-myhrvold-and-modernist-cuisine. 28 March 2011, Accessed 7 June 2013

Myhrvold N, Young C, Bilet M (2011) Modernist Cuisine: the art and science of cooking. 6 vols. The Cooking Lab, Bellevue, WA

Ott M, Choi Y, Cardie C, Hancock JT (2011) Finding deceptive opinion spam by any stretch of the imagination. In: Proceedings of the 49th Annual Meeting of the Association for Computational Linguistics. Association for Computational Linguistics. Stroudsburg, PA, 309–319

Park MY (2011) Nathan Myhrvold on Modernist Cuisine: The art and science of cooking. Epicurious.com.   http://www.epicurious.com/articlesguides/chefsexperts/interviews/nathan-myhrvoldinternview. Accessed 17 March 2012

Parsa A (2009) Exclusive: Belkin's development rep is hiring people to write fake positive amazon reviews. The Daily Background (blog). http://thedailybackground.com/2009/01/16/exclusive-belkins-development-rep-is-hiring-people-to-write-fake-positive-amazon-reviews.   16   Jan 2009, Accessed 7 June 2013

Potter J (2010) Cooking for geeks: real science, great hacks, and good food. O'Reilly Media, Sebastopol, CA

PR Newswire (2007) Cook's Illustrated reaches 1,000,000 in circulation. America's Test Kitchen. Press release. http://www.prnewswire.com/news-releases/cooks-illustrated-reaches-1000000-in-circulation-51599427.html. 21 March 2007, Accessed 7 June 2013

Puget Sound Business Journal (2012) Paying for Amazon.com Five-Star Reviews. http://www.bizjournals.com/seattle/morning_call/2012/01/paying-for-amazoncom-five-star-reviews.html. 27 Jan 2012, Accessed 7 June 2013

Reader's Digest Association (2012) Reader's Digest association completes sale of Allrecipes.com to Meredith. Press release. http://www.rda.com/news/readers-digest-association-completes-sale-of-allrecipes-com-to-meredith. 1 March 2012, Accessed 7 June 2013

Ruhlman M (2011) Cook from it? First, try lifting it. New York Times. http://www.nytimes.com/2011/03/09/dining/09modernist.html?pagewanted=all. 8 March 2011, Accessed 7 June 2013

Sagon C (2004) King of the kitchen. Washington Post. 18 Feb 2004

Silberman MS, Ross J, Irani L, Tomlinson B (2010) Sellers' problems in human computation markets. In: Proceedings of the ACM SIGKDD Workshop on Human Computation. ACM, New York, 18–21

Simonite T (2011) Hidden industry dupes social media users. MIT Technology Review. http://www.technologyreview.com/web/39304. 12 Dec 2011, Accessed 7 June 2013

Snyder MV (2008) Poison Study. Mira, Don Mills ON

Soulskill (2009) Belkin's Amazon rep paying for fake online reviews. Slashdot. http://hardware.slashdot.org/story/09/01/17/166226/belkins-amazon-rep-paying-for-fake-online-reviews.   17 Jan 2009, Accessed 7 June 2013

Stohs N (2009) These cooks have a lot on their plates. Milwaukee Journal Sentinel. http://www.jsonline.com/features/food/42598082.html. 7 April 2009, Accessed 7 June 2013

Streitfeld D (2011) In a race to out-rave, 5-star web reviews go for $5. New York Times. http://www.nytimes.com/2011/08/20/technology/finding-fake-reviews-online.html. 19 Aug 2011, Accessed 7 June 2013

Sun H (2010) Sellers' trust and continued use of online market places. Journal of the Association for Information Systems 11(4):182–211

Terdiman D (2006) Conde Nast working on its net. CNET. http://news.cnet.com/2100-1025_3-6107038.html. 18 Aug 2006, Accessed 7 June 2013

Toomey C (2010) What Ferran Adria is cooking up after El Bulli. London Sunday Times Magazine. http://www.christinetoomey.com/content/?p=392. 28 March 2010, Accessed 7 June 2013

Wang G, Wilson C, Zhao X, Zhu Y, Mohanlal M, Zheng H, Zhao BY (2012) Serf and turf: Crowdturfing for fun and profit. Prepublication version. http://arxiv.org/pdf/1111.5654v2.pdf. Accessed 3 Jan 2013

Index

W. Aspray et al., *Food in the Internet Age*, SpringerBriefs in Food,
Health, and Nutrition, DOI 10.1007/978-3-319-01598-9,
© William Aspray, George Royer, Melissa G. Ocepek 2013